An Assessment of Undersea Weapons Science and Technology

Committee for Undersea Weapons Science and Technology
Naval Studies Board
Commission on Physical Sciences, Mathematics, and Applications
National Research Council

NATIONAL ACADEMY PRESS
Washington, D.C.

NOTICE: The project that is the subject of this report was approved by the Governing Board of the National Research Council, whose members are drawn from the councils of the National Academy of Sciences, the National Academy of Engineering, and the Institute of Medicine. The members of the committee responsible for the report were chosen for their special competences and with regard for appropriate balance.

This work was performed under Department of the Navy Contract N00014-99-C-0307 issued by the Office of Naval Research under contract authority NR 201-124. However, the content does not necessarily reflect the position or the policy of the Department of the Navy or the government, and no official endorsement should be inferred.

International Standard Book Number 0-309-06926-2

Copies available from:

Naval Studies Board
National Research Council
2101 Constitution Avenue, N.W.
Washington, D.C. 20418

Printed in the United States of America

THE NATIONAL ACADEMIES

National Academy of Sciences
National Academy of Engineering
Institute of Medicine
National Research Council

The **National Academy of Sciences** is a private, nonprofit, self-perpetuating society of distinguished scholars engaged in scientific and engineering research, dedicated to the furtherance of science and technology and to their use for the general welfare. Upon the authority of the charter granted to it by the Congress in 1863, the Academy has a mandate that requires it to advise the federal government on scientific and technical matters. Dr. Bruce M. Alberts is president of the National Academy of Sciences.

The **National Academy of Engineering** was established in 1964, under the charter of the National Academy of Sciences, as a parallel organization of outstanding engineers. It is autonomous in its administration and in the selection of its members, sharing with the National Academy of Sciences the responsibility for advising the federal government. The National Academy of Engineering also sponsors engineering programs aimed at meeting national needs, encourages education and research, and recognizes the superior achievements of engineers. Dr. William A. Wulf is president of the National Academy of Engineering.

The **Institute of Medicine** was established in 1970 by the National Academy of Sciences to secure the services of eminent members of appropriate professions in the examination of policy matters pertaining to the health of the public. The Institute acts under the responsibility given to the National Academy of Sciences by its congressional charter to be an adviser to the federal government and, upon its own initiative, to identify issues of medical care, research, and education. Dr. Kenneth I. Shine is president of the Institute of Medicine.

The **National Research Council** was organized by the National Academy of Sciences in 1916 to associate the broad community of science and technology with the Academy's purposes of furthering knowledge and advising the federal government. Functioning in accordance with general policies determined by the Academy, the Council has become the principal operating agency of both the National Academy of Sciences and the National Academy of Engineering in providing services to the government, the public, and the scientific and engineering communities. The Council is administered jointly by both Academies and the Institute of Medicine. Dr. Bruce M. Alberts and Dr. William A. Wulf are chairman and vice chairman, respectively, of the National Research Council.

COMMITTEE FOR UNDERSEA WEAPONS SCIENCE AND TECHNOLOGY

ALBERT J. BACIOCCO, JR., The Baciocco Group, Inc., *Chair*
ARTHUR B. BAGGEROER, Massachusetts Institute of Technology
ALAN BERMAN, Center for Naval Analyses
GERALD A. CANN, Rockville, Maryland
A. DOUGLAS CARMICHAEL, Massachusetts Institute of Technology
THOMAS A. CLARE, Fredericksburg, Virginia
DENNIS F. COLIN, Lockheed Martin Federal Systems
ALEC D. GALLIMORE, University of Michigan
ERNEST L. HOLMBOE, Applied Physics Laboratory, Johns Hopkins University
ALFRED I. KAUFMAN, Institute for Defense Analyses
DAVID W. McCALL, Far Hills, New Jersey
L. DAVID MONTAGUE, Menlo Park, California
DOUGLAS R. MOOK, Sanders, a Lockheed Martin Company
RICHARD F. PITTENGER, Woods Hole Oceanographic Institution
DANIEL SAVITSKY, Stevens Institute of Technology

Staff

CHARLES F. DRAPER, Study Director
SIDNEY G. REED, JR., Consultant

Navy Liaison Representative

SPYRIDON G. LEKOUDIS, Acting Head, Engineering, Materials and Physical Science and Technology Department, Office of Naval Research

NAVAL STUDIES BOARD

VINCENT VITTO, Charles S. Draper Laboratory, Inc., *Chair*
JOSEPH B. REAGAN, Saratoga, California, *Vice Chair*
DAVID R. HEEBNER, McLean, Virginia, *Past Chair*
ALBERT J. BACIOCCO, JR., The Baciocco Group, Inc.
ARTHUR B. BAGGEROER, Massachusetts Institute of Technology
ALAN BERMAN, Applied Research Laboratory, Pennsylvania State University
NORMAN E. BETAQUE, Logistics Management Institute
JAMES P. BROOKS, Litton/Ingalls Shipbuilding, Inc.
NORVAL L. BROOME, Mitre Corporation
JOHN D. CHRISTIE, Logistics Management Institute
RUTH A. DAVID, Analytic Services, Inc.
PAUL K. DAVIS, RAND and the RAND Graduate School
SEYMOUR J. DEITCHMAN, Chevy Chase, Maryland, *Special Advisor*
DANIEL E. HASTINGS, Massachusetts Institute of Technology
FRANK A. HORRIGAN, Bedford, Massachusetts
RICHARD J. IVANETICH, Institute for Defense Analyses
MIRIAM E. JOHN, Sandia National Laboratories
ANNETTE J. KRYGIEL, Great Falls, Virginia
ROBERT B. OAKLEY, National Defense University
HARRISON SHULL, Monterey, California
JAMES M. SINNETT, Boeing Company
WILLIAM D. SMITH, Fayetteville, Pennsylvania
PAUL K. VAN RIPER, Williamsburg, Virginia
VERENA S. VOMASTIC, The Aerospace Corporation
BRUCE WALD, Center for Naval Analyses
MITZI M. WERTHEIM, Center for Naval Analyses

Navy Liaison Representatives

RADM RAYMOND C. SMITH, USN, Office of the Chief of Naval Operations, N81
RADM PAUL G. GAFFNEY II, USN, Office of the Chief of Naval Operations, N91

Marine Corps Liaison Representative

LTGEN JOHN E. RHODES, USMC, Commanding General, Marine Corps Combat Development Command

RONALD D. TAYLOR, Director
CHARLES F. DRAPER, Senior Program Officer
MARY G. GORDON, Information Officer
SUSAN G. CAMPBELL, Administrative Assistant
JAMES E. MACIEJEWSKI, Senior Project Assistant

Preface

The Department of the Navy strives to maintain, through its Office of Naval Research (ONR), a vigorous science and technology (S&T) program in those areas considered critically important to U.S. naval superiority in the maritime environment, including littoral waters and shore regions. In pursuing its S&T investments in such areas, ONR must ensure that (1) a robust U.S. research capability to work on long-term S&T problems in areas of interest to the Department of the Navy and the Department of Defense is sustained, (2) an adequate supply of new scientists and engineers in these areas is maintained, and (3) S&T products and processes necessary to ensure future superiority in naval warfare are provided.

One of the critical areas for the Department of the Navy is undersea weapons. At the request of ONR, the National Research Council established the Committee for Undersea Weapons Science and Technology, under the auspices of the Naval Studies Board, to assess the S&T in this area.

The terms of reference of the study called for (1) assessing the health of the existing Navy program in undersea weapons, (2) evaluating the Navy's research effort to develop the capabilities needed for future undersea weapons, (3) identifying non-Navy-sponsored research and development efforts that might facilitate the development of such advanced weapons capabilities, and (4) recommending how the Navy's research program should be focused so that it can meet future needs. In addition, the existing program was to be assessed in the following areas:

- Maturity of the key technology areas and challenges and cost drivers in those areas;
- Interaction with related technology areas;
- Program funding and funding trends;
- Scope of naval responsibility;
- Scope, degree, and stability of non-Navy activities in key technology areas;

- Performer base (academia, government, industry, foreign);
- Infrastructure (leadership in the area);
- Knowledge-base pipeline (graduate, postdoctoral, and career delineation);
- Facilities and equipment (ships, test tanks, and the like); and
- Integration with and/or transition to higher-budget-category programs.

Finally, the assessment would answer two key questions: (1) What technology developments that are not being addressed, or that are being addressed inadequately, are needed to meet the Navy's long-term objectives? and (2) To what extent does the development of these technologies depend on Navy-sponsored research and development?

The committee was composed of individuals from a variety of backgrounds and organizations. At its initial meeting, the committee received extensive briefings on the aims and accomplishments of the ONR undersea weapons S&T program (housed primarily in Code 333); this information was supplemented by additional information obtained through individual discussions with research performers and experts in the field. The committee's subsequent deliberations on the existing program and its adequacy were based on information provided in those briefings and discussions. Furthermore, in the report's discussion, findings, conclusions, and recommendations, the committee does not undertake to suggest a detailed restructuring of the program, but indicates the broad directions it believes the ONR program should take.

The study began in August 1999 and lasted for approximately 5 months. During that time, the committee held five meetings:

- August 30-31, 1999, in Washington, D.C. Organizational meeting with briefings provided by ONR, the Program Executive Office for Undersea Warfare, the Naval Undersea Warfare Center, and the Applied Research Laboratory at Pennsylvania State University.
- September 14-15, 1999, in Washington, D.C. Briefings were provided by ONR, the Office of Naval Intelligence, the Naval Surface Warfare Center, and the Defense Advanced Research Projects Agency.
- October 18-19, 1999, in Washington, D.C. Briefings were provided by ONR and the Office of the Chief of Naval Operations [Anti-Submarine Warfare Requirements Division (N84), Surface Warfare Division (N86), and Submarine Warfare Division (N87)].
- November 9-11, 1999, in Washington, D.C. Briefings were provided by the Office of the Chief of Naval Operations [Air Warfare Division (N88) and Expeditionary Warfare Division (N85)], the Defense Intelligence Agency, the Naval Research Advisory Committee, the Naval Undersea Warfare Center, and the Applied Research Laboratory at Pennsylvania State University.
- December 7-8, 1999, in Washington, D.C. Briefings by ONR.

In addition, on November 3, 1999, two committee members visited the Applied Research Laboratory at Pennsylvania State University in State College, Pennsylvania.

The resulting report, prepared in the ensuing several months, represents the committee's consensus view on the issues raised and questions posed in the terms of reference.

Acknowledgment of Reviewers

This report has been reviewed by individuals chosen for their diverse perspectives and technical expertise, in accordance with procedures approved by the National Research Council's (NRC's) Report Review Committee. The purpose of this independent review is to provide candid and critical comments that will assist the authors and the NRC in making the published report as sound as possible and to ensure that the report meets institutional standards for objectivity, evidence, and responsiveness to the study charge. The contents of the review comments and the draft manuscript remain confidential to protect the integrity of the deliberative process. The committee wishes to thank the following individuals for their participation in the review of this report:

William H. Arnold, Westinghouse Electric Corporation (retired),
Shirley Chiang, University of California at Davis,
VADM James Fitzgerald, USN (retired), Analysis and Technology, Incorporated,
Paul A. Fleury, University of New Mexico,
RADM Richard Riddell, USN (retired), General Dynamics,
Jack M. Sipress, Sipress Associates, and
Edward Zdankiewicz, Arnold, Maryland.

Although the individuals listed above provided many constructive comments and suggestions, responsibility for the final content of this report rests solely with the authoring committee and the NRC.

Contents

Appendixes

Executive Summary

From a broad perspective, undersea weapons science and technology (S&T) is the result of fundamental and applied research and provides the basis for producing any of the Navy's weapons that spend some time undersea. This arsenal includes, for example, mine-like torpedoes, submarine-launched mobile mines, submarine-launched missiles, and, most important for this study, torpedoes and torpedo countermeasures.

APPROACH AND CONTENT

The Committee for Undersea Weapons Science and Technology developed its study in response to the terms of reference outlined in the Preface. Chapter 1 discusses the Office of Naval Research's (ONR's) role in and responsibility for naval S&T. Chapter 2 presents the committee's assessment of the ONR undersea weapons S&T program and the uniqueness of the program's responsibility vis-à-vis the Navy. Since the ONR undersea weapons S&T program is predominantly torpedo-related, the assessment follows the torpedo program categories: warheads, propulsion, guidance and control, torpedo stealth, torpedo defense, countermeasures, supercavitating weapons, and weapons design optimization. In the "Summary of Assessment" in Chapter 2, the committee also addresses the two key questions posed in the terms of reference. In Chapter 3, the committee presents its perspectives on major issues surrounding the future of Navy undersea weapons construed broadly (i.e., including but not limited to torpedoes), on ways to approach these issues, and on the implications for Navy S&T. Findings, conclusions, and recommendations are presented in Chapter 4.

ASSESSMENT OF ONR UNDERSEA WEAPONS S&T PROGRAM

The committee offers the following assessments of the ONR undersea weapons S&T program:

- Torpedo upgrades are mature.
- Interactions with related technology areas need strengthening.

- Program funding is not sufficient to offset the evolving S&T available to potential enemies. In particular, basic research funding (Department of Defense budget category 6.1) is much too small.
- Because they must function in a challenging undersea environment, undersea weapons involve special technologies, adaptations of other technologies, and unique integration of all these technologies. There is in the United States no sustained non-Navy support for this type of effort.
- The U.S. Navy must make a greater effort to provide leadership in undersea weapons research and development if it wishes to match the activity and capability of other nations.
- The knowledge-base pipeline is adequate to support the current program, although undersea weapons research is not viewed as a particularly attractive career path. However, this pipeline would be hard pressed to support the level of activity required for the development of next-generation weapon systems, which will be increasingly sophisticated in virtually all the critical technology areas.
- Facilities and equipment are not in short supply, although distributed simulation facilities in greater numbers and capability will be needed.
- The integration of the ONR undersea weapons S&T program with torpedo programs in higher-order budget categories is too tight. Basic and applied research that could lead to revolutionary weapons efforts is being neglected.

In answer to the first key question in the terms of reference, which asked what technologies are needed but are not being developed by the ONR undersea weapons S&T program, the committee offers the following judgments:

- The current approach to effectively confronting submarines in the littoral environment is not founded on a complete analysis and a good understanding of the physics of the problem, and it needs attention at the most basic level. Within the ONR undersea weapons S&T program, support for the underlying S&T is minimal.
- Deployable, distributed sensor arrays are a promising technology that needs to be built upon, as does related work in data fusion and undersea communications.
- Undersea weapons signal processing applications of fiber-optic bandwidth need to be exploited.
- Unmanned underwater vehicles and small manned underwater vehicles could be employed by naval forces as semiautonomous, long-endurance hunter/killers and reconnaissance vehicles.
- Alternative prime power concepts (e.g., hybrid advanced electric and internal combustion systems) that might be applicable to weapon-carrying and reconnaissance undersea vehicles need to be part of the exploratory program.
- Within the ONR undersea weapons S&T program, the committee received no indication of program activity in short-action-time, rocket-propelled, air- or surface-delivered undersea weapons.
- The committee did not note any programs based on other than traditional torpedo concepts.
- There is a need for the disciplined use of operations and systems analysis as a means to evaluate, quantify, and guide program decisions.

In answer to the second key question in the terms of reference, namely, the extent to which undersea weapons S&T depends on Navy-sponsored research and development (R&D), the committee believes that the U.S. Navy bears the main responsibility for such S&T because undersea weapons must function in a challenging undersea environment and they involve special technologies, adaptations of other technologies, and unique integration of all these technologies. There is in the United States no sustained support for this type of effort outside the Navy.

FINDINGS, CONCLUSIONS, AND RECOMMENDATIONS

Based on its assessment of the ONR undersea weapons S&T program, the committee offers the following findings, conclusions, and recommendations.

Finding 1. Undersea weapons involve special technologies, adaptations of other technologies, and unique integration of all these technologies for which there is in the United States little non-Navy-sponsored research and development. The committee found some good examples of ongoing healthy and productive S&T, including the following:

- The program on propulsion at the Applied Research Laboratory, Pennsylvania State University (ARL/PSU) is exemplary and offers technologies for both weapons and vehicles that could be used in future systems. Closed-cycle engines are among the increasingly attractive options as the importance of stealth and endurance increases.
- The programs on warheads at the Naval Surface Warfare Center, Indian Head are good examples of R&D in a mature area that has consistently delivered fresh results in S&T and new generations of explosive compounds tailored to the Navy's needs. Current research on the penetration of hardened hulls is important. Research on the problems of sensitivity of high-energy materials should be supported.
- The program at ARL/PSU and the Naval Undersea Warfare Center (NUWC) to develop a high-speed, supercavitating vehicle is challenging and sufficiently promising to warrant research in (1) the physics of supercavitating flow, (2) vehicle control and guidance methodology, and (3) the design and building of a testbed. There should be careful analysis of the operational utility of the concepts this technology could include.

On the other hand, the committee believes that a truly healthy undersea weapons S&T program should include industry participation, but industry is not now a significant participant or investor in undersea weapons S&T.

Conclusion 1. The Navy has a unique responsibility for the support and health of S&T related to undersea weapons construed broadly. Although the information presented to the committee pertained only to torpedo-related matters, the committee believes that its recommendation on this responsibility should apply to all weapons spending some time undersea. The undersea weapons S&T effort should include industrial participation, at least in relevant concept definition studies, and related operations and systems analysis.

Recommendation 1. The Navy should designate S&T for undersea weapons—construed broadly—as a National Naval Need.[1] Because of the key enabling characteristics of undersea weapons for the fleet and the need for industry involvement, the Navy should also consider designating undersea weapons as one of the Future Naval Capabilities (FNCs),[2] a step that would allow it to begin preparation for a new weapons acquisition program.

---

[1]As stated by Fred E. Saalfeld to the Office of Naval Research (ONR), National Naval Programs (now called National Naval Needs) are those science and technology areas that are uniquely important to the naval forces and whose health depends on ONR investment.

[2]Future Naval Capabilities are chosen by a top-level Navy and Marine Corps board, and the corresponding S&T is defined in detail by integrated product teams. See further explanation under "Strategy of the Office of Naval Research" in Chapter 1.

Finding 2. There is no broadly based, future-oriented program of operations and systems analysis in place to support ONR S&T planning in undersea weapons. With regard to the individual ONR undersea weapons program areas, the committee found the following:

- It is not obvious that the programs on guidance and control at ARL/PSU and NUWC are succeeding at coping with progressively quieter targets and evolving countermeasures. The careful operations and systems analysis needed to critically assess operational performance in matters of target detection, identification, and homing seems to be missing.
- Upgrades intended to quiet the MK-48 and MK-54 torpedoes (mainly by NUWC) were not persuasively presented to the committee. The open-cycle engine, buoyancy disadvantages, hydroacoustic noise, and other characteristics make the upgrades questionable in light of the evolving stealth and countermeasure capabilities of potential enemy targets. No systems analyses of predicted program success or time scales for acquisition were presented to the committee.
- A number of plausible approaches to defending against torpedoes were broadly outlined to the committee, including noisemakers, decoys, supercavitating pellets, and antitorpedo torpedoes. Individually these might be of value, but maximum benefit will be achieved only if they are integrated properly into a plausible, coherent defense architecture system.
- Weapons design optimization, which appears to be a relatively recently identified effort, while useful still does not satisfy the need for operations and systems analysis called for at several points in this report.

Conclusion 2. Concept definitions, and systems and operational analysis, are needed in a number of program areas and as a part of a healthy and productive S&T process generally.

Recommendation 2. ONR should rigorously implement a process of operations and systems analysis of undersea weapons systems. Operational performance in both littoral and blue water environments should be covered. Emphasis should be placed on enabling science and technology and weapons systems of advanced mission and design.

Finding 3. The health of the existing Navy program on undersea weapons S&T is strongly affected by the present emphasis on upgrades of existing torpedo systems. Less than 10 percent of the ONR undersea weapons S&T budget was for basic research (6.1) in 1999. The health of the program could be improved by much greater attention to S&T issues that will affect future weapons systems.

Conclusion 3. To be more forward-looking, greater S&T emphasis at a fundamental (6.1/6.2) level is needed within the ONR undersea weapons effort.

Recommendation 3. ONR should increase undersea weapons S&T funding sufficiently to satisfy Future Naval Capability goals while ensuring that longer-range, higher-risk/higher-payoff alternative S&T is also enabled. This investment in future systems must be protected against raids to bail out near-term projects.

Finding 4. While some of the items covered in the present program, which is focused on upgrades of existing torpedoes, may also be useful in future weapons systems (e.g., propulsion units and warheads), long-range exploration for fundamentally new undersea weapons concepts and missions is needed.

Conclusion 4. Innovation beyond current undersea weapons concepts and missions is needed for a healthy S&T program. S&T should be pursued toward torpedoes operating with sensor arrays, unmanned and manned undersea vehicles (attack, reconnaissance, and so forth), and sophisticated mines, and toward achievement of short-action-time air-delivered undersea weapons systems. In the future, undersea weapons systems will be driven increasingly by overall architecture that will demand much more interdisciplinary coordination than was seen by the committee in the course of its study.

Recommendation 4. ONR should take a broader and longer-range view of undersea weapons systems, specifically those not limited to torpedoes and beyond the 5-year horizon. Some such concepts are being explored in other ONR undersea technology activities and in the Defense Advanced Research Projects Agency and should be considered.

1

Introduction

The challenge of undersea warfare is not widely appreciated. In the atmosphere, objects are generally visible, locatable, and identifiable. Electromagnetic radiation is reflected, refracted, and absorbed to a manageable extent. The boundaries do not cause major problems. The movement of platforms and projectiles is not greatly impeded by the medium. Electromagnetic emissions from sources occur at the speed of light, resulting in high data rates, which facilitate detection, classification, localization, and engagement.

By comparison, the underwater challenge in these respects is limited by the speed of sound in seawater. Radiation is strongly absorbed and redirected at boundaries. Deep water provides cover. Acoustic methodology has long been of paramount importance in deep water, and although the environment is relatively well understood, it is complicated. Near-shore waters (the littorals) are essential battle spaces for extending a force, but the challenges are enormous and not well characterized in terms of science or terrain. Uncertainty is common. Acoustic signals respond to all manner of (often unknown) boundaries associated with shallow water, traffic, debris, gradients, and the like. The encounter distances may be short. Water absorbs radiation directly and is often murky besides.

The physics of the undersea environment is probably understood in general terms, but coupling this understanding to practical use is difficult. New physics and new applications will emerge to the benefit of nations that pursue the basic science. A broadening of sensor technology is inevitable. Chemistry will be exploited. It is in this domain of open-ended science and technology (S&T) that the Office of Naval Research (ONR) can make a unique contribution by identifying, initiating, and enabling future opportunities, missions, capabilities, and solutions.

The Department of the Navy's undersea weapons S&T program resides at ONR and addresses primarily torpedo-related issues. The Department's inventory of weapons that spend some time undersea includes mine-like torpedoes (e.g., fixed and mobile mines such as the CAPTOR mine, which is an encapsulated torpedo), the submarine-launched mobile mine, and the ASROC antisubmarine rocket-propelled torpedo launched vertically from a surface ship, as well as strike weapons such as the Tomahawk missile, launched from a submerged submarine. At the sponsor's request and with the concurrence of the Committee for Undersea Weapons Science and Technology, this assessment focused on that

Box 1.1 Approximate 20-Year Period for Transition to New-technology Torpedo

Change from MK 37 to MK 48

- Antisubmarine warfare was the top Navy priority
- Full production capabilities
- Experienced personnel
  - Design and develop operations evaluation and initial operating capability: ~9 years
  - Solve reliability problems: ~4 years
  - Develop fire control modifications and train people: ~7 years

Change from MK 46 to MK 50

- Development started in 1972
- FY89 Secretary of Defense Annual Report stated, "We now anticipate cost increases and a 21-month delay in MK-50 full-scale development program."
- MK 50 finally passed operation evaluation OTIIB in 1992.

SOURCE: John Zittel, "Undersea Weapons S&T," Office of the Chief of Naval Operations, N84T, Washington, D.C., briefing to the committee, October 18, 1999.

part of ONR's undersea weapons S&T program that does not include any of these other kinds of weapons but that includes primarily torpedoes and torpedo countermeasures. Nevertheless, in its discussions of the Department of the Navy's future needs and responsibilities for undersea weapons, the committee believed that a broader perspective was needed, including not only torpedoes but also mines and other weapons that spend some time undersea.

The Navy's current inventory of torpedoes includes the heavyweight (submarine-launched) MK 48 and the lightweight (ship- or air-launched) MK 46 and the newer MK 50. These are based on designs and upgrades over the past 30 to 50 years. Historically, torpedoes have been developed on a 15- to 20-year cycle, as shown in Box 1.1. The MK-48 heavyweight can be used in antisubmarine warfare (ASW) and in antisurface warfare; the lightweight MK 46 and MK 50 can be used primarily for ASW. The MK-54 lightweight hybrid torpedo has a faster developmental schedule and is low cost.

With the end of the Cold War, in the early 1990s, the focus of naval strategy moved to the littorals. In littoral waters, depths can vary considerably. In deeper littoral waters, the Cold War-type challenges of quiet submarines and torpedo counter-countermeasures remain. In shallow littoral waters, the ASW problem is even more complex and difficult because of the available countermeasures, the environment, and the stealthiness and small size of undersea diesel-electric submarines, all of which challenge the performance of the Navy's undersea detection and weapons systems.

During this same post-Cold War period, Navy funding for torpedoes has been drastically reduced, as shown in Figure 1.1. While the overall Navy budget has dropped some 33 percent and its acquisition budget 42 percent,[1] the budget for torpedoes has gone down by a factor of 7. There is currently no U.S.

[1]Naval Studies Board, National Research Council. 1997. *Recapitalizing the Navy: A Strategy for Managing the Infrastructure,* National Academy Press, Washington, D.C., p. 11.

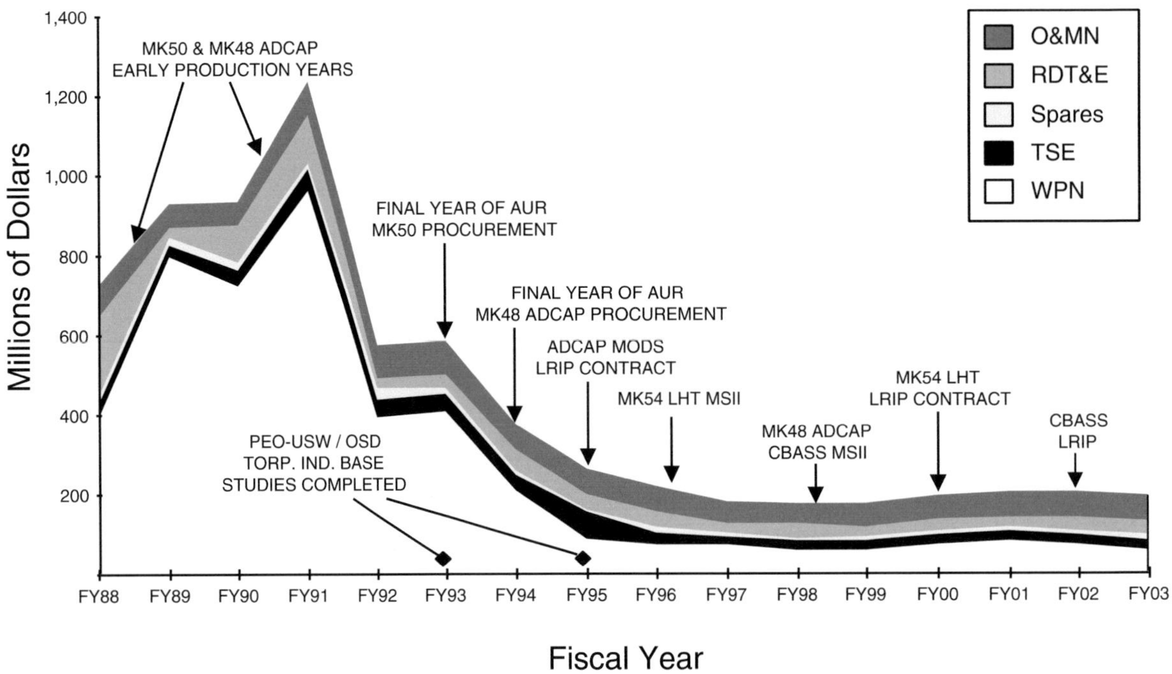

FIGURE 1.1 Program funding plan and trends—total torpedo and vertical launch antisubmarine rocket (VLA) funding, FY88-FY03. This graph illustrates that undersea weapons funding has been cut disproportionately during the past decade. Acronyms are defined in Appendix D.
SOURCE: John Zittel, "Undersea Weapons S&T," Office of the Chief of Naval Operations, N84T, Washington, D.C., briefing to the committee, October 18, 1999.

torpedo production (other than the Mk-54 LRIP). Foreign countries, however, are actively producing increasingly sophisticated torpedoes. To meet these new challenges, future undersea weapons will need to have significantly improved characteristics, including reduced size to permit more weapons on platforms; reduced acquisition and life-cycle costs; longer shelf life; better deep- and shallow-water and counter-countermeasures performance; greater endurance; flexible speed control; stealth; lethality; and, for urgent-attack weapons, shorter reaction times. Technology advances toward these improved characteristics are grouped into the following areas: undersea warheads and explosives; energy conversion and propulsion; guidance and control; hydrodynamics, especially of high-speed supercavitating weapons; undersea warheads and explosives; simulation and testing; modular systems integration; and ship torpedo defense systems. A 1996 Department of the Navy master plan for undersea weapons, vehicles, and countermeasures[2] provides a road map for torpedo-related technology developments and a schedule of technology insertions (see Appendix A).

[2]Department of the Navy. 1996. *Undersea Weapons, Vehicles, and Countermeasures: Master Plan*, The Pentagon, Washington, D.C.

MISSION OF THE OFFICE OF NAVAL RESEARCH

ONR has been a widely respected sponsor of basic research (6.1) for more than 50 years and has added almost all Department of the Navy applied research and advanced development (6.2 and 6.3) to its responsibilities for the last decade. F.E. Saalfeld, Executive Director and Technical Director of ONR, described ONR's function in the following terms:

> ONR is in the business of investing in science and technology. And the return on that investment isn't in the first instance profit, as it would be with industry, or even knowledge, as it would be with academia. The ultimate return on our science and technology investment is warfighter capabilities. . . . As the manager of the Department of the Navy Science and Technology Program, the Office of Naval Research will continue to ensure that the portfolio includes the best available mix of investment partners and research performers. And since our ultimate shareholders are sailors and Marines, the return on investment we look for in Naval science and technology is not profits, but capabilities.[3]

He described the mission of ONR as follows:

> ONR is the "front-end" of the acquisition pipeline. Its mission is to provide the S&T base that maintains and expands the technological superiority of naval forces. Its goal is to respond to naval requirements and lead the international S&T community, to provide both evolutionary technology improvements and revolutionary capabilities.
>
> The Navy and Marine Corps operate on—and above, and under, and from—the sea. The maritime environment extends from the sea floor to earth orbit. It is complex and challenging, and it makes Naval operations inherently difficult and dangerous even under the best conditions. The Department of the Navy has therefore historically placed great emphasis on maintaining a vigorous science and technology program in those areas where research is critically important to maintaining United States Naval superiority.
>
> A lot of those areas, uniquely important to the Navy and Marine Corps, are simply not addressed by research investments from the other Services, or for that matter from the National Science Foundation, the National Institutes of Health, other federal research establishments, or even private industry. This means that the health, strength, and growth of our scientific and technical capabilities in those fields depend upon the Department of the Navy.
>
> On behalf of the Department of the Navy, the Office of Naval Research must ensure continuing United States leadership in these vitally important scientific and technical disciplines. It does so through research, recruitment, and education, all done with a view to sustaining an adequate base of talent and the critical infrastructure necessary to carry out research and experimentation.[4]

It is with these expectations in mind that the committee undertook the review called for in the terms of reference (see Preface).

---

[3]Saalfeld, Fred. E., Office of Naval Research. Speech given at the Naval Science Assistance Program (NSAP) All-Theater Conference, held at the Naval Research Laboratory, Washington, D.C., June 23, 1999.

[4]Saalfeld, Fred E. 2000. *ONR Presents . . . National Naval Responsibilities in Science and Technology,* Office of Naval Research, Arlington, Va., February 27.

STRATEGY OF THE OFFICE OF NAVAL RESEARCH

The committee was given information on ONR's current strategy for S&T. The strategy includes a continuation of fundamental research (6.1), applied research (6.2), and advanced technology development (6.3). To facilitate the transition to efforts in higher budget categories, about half of the 6.2 and most of the 6.3 effort will be concentrated in areas corresponding to budgetary spikes supporting Future Naval Capabilities (FNCs).[5] The FNCs are chosen by a top-level Navy and Marine Corps board, and the corresponding S&T is defined in detail by integrated product teams. The spikes are to have approximately 3- to 5-year lifetimes and include scheduled milestones.

The current ONR undersea weapons S&T program includes elements that support several of the FNCs, notably littoral antisubmarine warfare and platform protection.[6] The ONR strategy also identifies S&T programs (mainly 6.1 and 6.2) that address National Naval Needs, which constitute a national naval responsibility. It is understood that National Naval Needs should have continuity and a stable budget.

The processes for executing this strategy are under way. While the undersea weapons S&T program has not yet been identified as a National Naval Need, a major motive for the present study is to examine whether it should be so identified.

REPORT OUTLINE

Chapter 2 of this report contains the committee's assessment of the existing ONR undersea weapons S&T program, following the aforementioned torpedo-related categories into which the program's budget is divided. The chapter also reviews the program from the point of view of the issues listed in the terms of reference and attempts to answer the two questions posed there. Chapter 3 contains the committee's views on the future of the Navy's undersea weapons in a broad context not limited to torpedoes. Findings, conclusions, and recommendations are presented in Chapter 4.

---

[5]RADM Paul G. Gaffney II, USN, Director, Test and Evaluation and Technology Requirements. Future Naval Capabilities Fiscal Guidance – Information Memorandum, Office of the Chief of Naval Operations (N91), The Pentagon, Washington, D.C., November 23, 1999.

[6]Outside the S&T program, platform-related acquisition programs, e.g., submarines and DD-21, can also affect the Navy undersea warfare funding structure.

2

Assessment of the Office of Naval Research's Undersea Weapons Science and Technology Program

This chapter presents the committee's assessment of the existing Office of Naval Research (ONR) undersea weapons S&T program, area by area: warheads, propulsion, guidance and control, torpedo stealth, torpedo defense, supercavitating weapons, and weapons design optimization. The questions posed in the statement of task are then addressed. Planned 6.2 and 6.3 (i.e., BA2 and BA3) funding for the ONR program is shown in Figure 2.1. Figure 2.2 depicts the funding history of the ONR undersea weapons program, including, since FY94, some 6.1 and other (e.g., congressional "plus ups") funding.

UNDERSEA WEAPONS TECHNOLOGIES

Warheads

In the past, the thrust of S&T for underwater explosives was to achieve an increased explosive yield from warheads of the same weight. As the vehicles that are the targets of the torpedoes become smaller and are presumed to be less robust, the new thrust is to achieve the same explosive lethality with a smaller, lighter warhead. Lighter warheads with constant lethality would reduce the negative buoyancy of the torpedoes and permit them to run at lower speeds. Lower speeds would reduce the radiated noise of a torpedo, enhancing its sonar performance and reducing the target's time to detect and, if necessary, identify an attack. Lower speeds might also enable longer run times, permitting the torpedo to loiter or search as well as attack.

Since the explosive yield of an underwater warhead varies with the explosive weight W as $W^{1/3}$, significant increases in explosive yields must be achieved before the weight of the warhead can be significantly reduced. Certainly more energetic materials are known than those employed in current torpedoes. Unfortunately, the higher the energy, the more unstable the compound tends to be. One of the principal directions of explosive materials research has been the formulation of insensitive warheads that provide both high yield and operational safety.

The search for more effective explosives has usually followed the synthesis of chemicals that have a high energy density. Owing to the inherent stability of nitrogen gas (N_2), which is a product of the

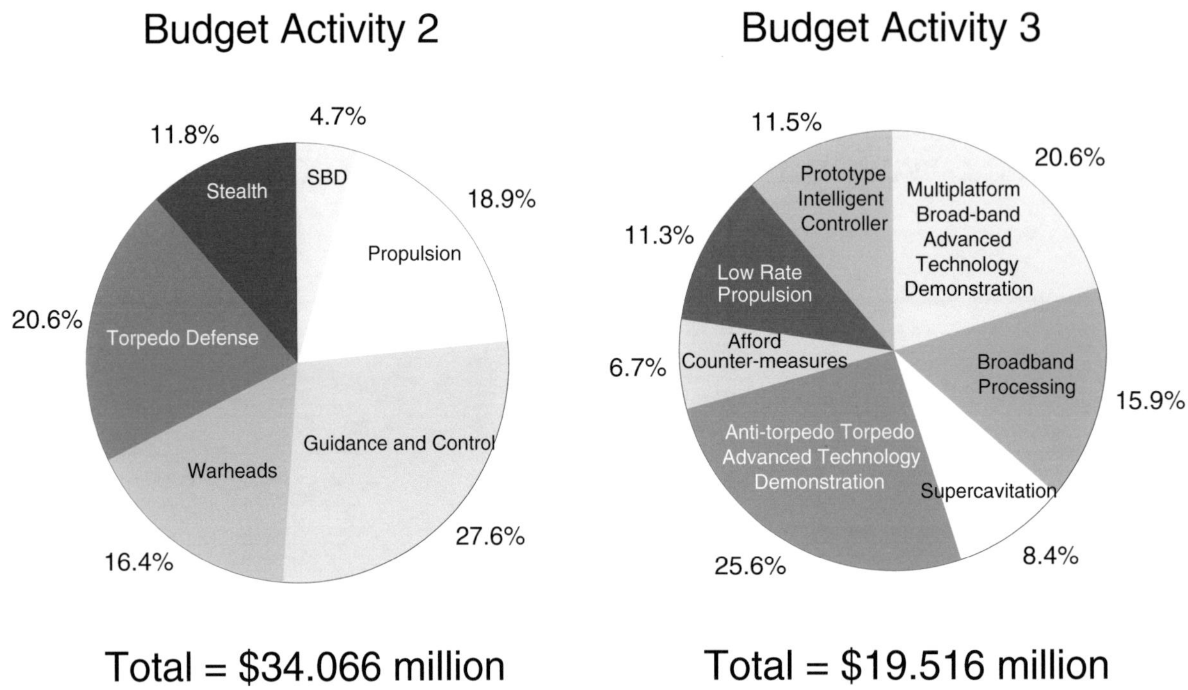

FIGURE 2.1 Undersea weaponry thrust as indicated by planned FY00 resources for Department of the Navy S&T, which is listed by area. SBD, simulation-based design.
SOURCE: Spyridon G. Lekoudis, "Undersea Weaponry Thrust," Office of Naval Research, Arlington, Va., briefing to the committee, Washington, D.C., August 30, 1999.

oxidation of nitrogen-containing compounds, nitration has been a successful approach to the synthesis of new generations of more effective explosives. The oxygen in the nitrate group (NO_3) provides the oxidant for the carbon and hydrogen. A compound called RDX is widely used in Navy warheads. A compound designated as HMX is about 10 percent more energy dense and has been superseding RDX when safe applications can be found. There are indications that HMX may be nearing the upper limit of energy density that can be made safe from inadvertent explosion, i.e., made safely "insensitive." The Navy has had unfortunate accidents that have led to the policy that all new explosives must be proved safe. A new and promising explosive, CL-20, is now in the process of being certified as insensitive.

There are other families of chemicals that are not based on the nitration of organic compounds. Perchlorates (ClO_4^-), for instance, are also high explosives, and their balance of energy density and sensitivity can be favorable. ONR has supported research and development on perchlorates, and some formulations are in use.

Undersea explosives intended to destroy the hulls of ships and submarines often include aluminum in order to produce a large bubble, the dynamics of which in turn can place a large stress on the metal structure under attack. The advantage of using aluminum is partially lost by the passivation of the surface of the aluminum particle in the explosive by an inert layer of Al_2O_3, which leaves unreacted aluminum under the surface layer. To overcome this defect, ONR has an S&T program to replace some nitro- or nitrate groups with NF_2. Doing so enhances the energy release, since the fluorine tends to undermine the Al_2O_3 and increases the activity of the aluminum by causing it to react more completely.

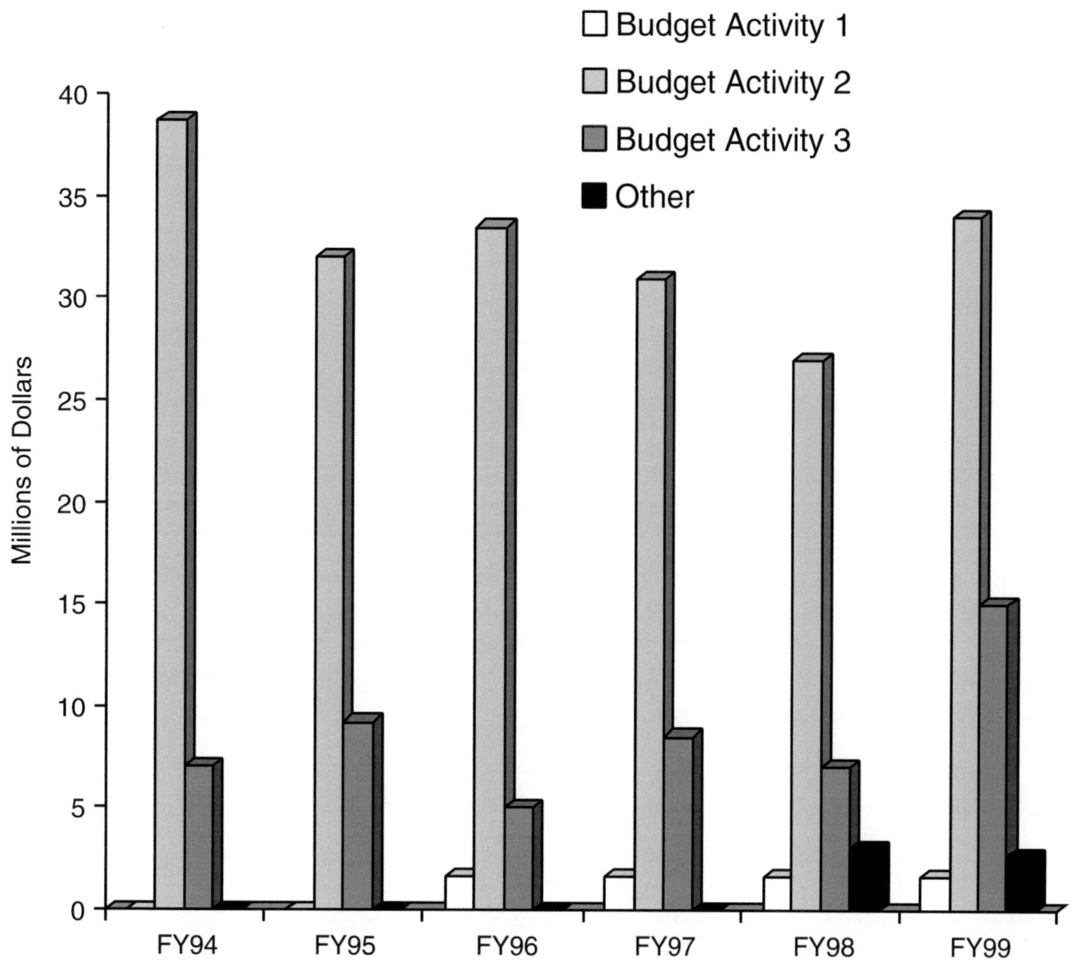

FIGURE 2.2 Department of the Navy undersea weaponry thrust, FY94 through FY99.
SOURCE: Spyridon G. Lekoudis, "Undersea Weaponry Thrust," Office of Naval Research, Arlington, Va., briefing to the committee, Washington, D.C., August 30, 1999.

This approach, however, may also sensitize the material, so a sufficient but partial insertion of NF_2 appears to be necessary and is being investigated.

ONR has a long record of supporting studies of insensitive munitions. Formulations with various plastic materials have been highly successful. Unintended detonations are probably initiated by abrasion of the explosive particles. Formulations with nonexplosive binders (usually commercial plastics) can reduce sensitivity at the cost of dilution and lost energy density. ONR sponsors efforts to find optimum compositions. For example, the use of reactive binders (e.g., cellulose nitrate or NF_2-rich plastics) is an approach that might be effective.

ONR is supporting research into the origins of insensitivity. Good progress is being made using techniques that can characterize chemical reactions on very short time scales. This work could lead to strategies for the synthesis of new families of more energy-dense but safe materials.

ONR also supports the development of warheads with an improved shaped charge. These warheads would be better able to penetrate a hull by concentrating their explosive energy over a relatively small area of the target hull. Shaped charges can significantly reduce the amount of explosive needed to achieve the desired level of lethality. As pointed out above, a reduction in warhead weight might be used to enable improvements in other important weapon performance characteristics.

Under ONR sponsorship, progress has been achieved in the safety, arming, and fusing of undersea weapons. With the use of advanced commercial electronics and microelectromechanical systems, the volume of the exploder has been reduced by more than 80 percent. While a reduction in the size of the fuse is important in all explosive-carrying undersea weapons, it is essential to the development of the new 6.25-in. antitorpedo torpedoes.

Modeling of warhead-target interactions is another important area of study that ONR sponsors. These models are important adjuncts to experimental programs, as they will enhance understanding of the physics of the process, reduce the costs of development, and provide estimates of warhead lethality in situations that are difficult to test.

The committee noted that ONR research on explosives should be applicable also to sea mines. No example of this aspect of the research was presented.

Finding: The programs on warheads at the Naval Surface Warfare Center, Indian Head are good examples of research and development (R&D) in a mature area that has consistently delivered fresh results in S&T and new generations of explosive compounds tailored to the Navy's needs. Current research on the penetration of hardened hulls is important. Research on the problems of sensitivity of high-energy materials should be supported.

Propulsion

Much of the volume and weight of torpedoes is usually taken up by the energy and propulsion systems, so it is obviously important to emphasize the S&T base of these systems. In addition, the energy and propulsion systems are usually the noisiest components in most torpedoes.

Current torpedoes use the monopropellant OTTO-fuel II for MK-46 and MK-48 torpedoes and the Stored Chemical Energy Propulsion System (SCEPS) for the MK-50 torpedo. The OTTO-fuel II system operates as an open cycle and discharges the exhaust to the ocean, while the SCEPS system has a closed-cycle engine, a constant overall weight, higher energy density, and less radiated noise.

The absence of atmospheric oxygen provides a challenge for many energy systems in undersea vehicles. Other oxidants must be used, such as sulfur hexafluoride (SF_6) in the SCEPS unit, and there is only limited experience with such systems. Reliability and safety are important concerns for these energy sources because of their novelty and because these systems usually involve very energetic reactions. Cost and the difficulty of maintenance with SCEPS were cited as reasons for replacing the MK-50 engine with that from the MK 46 in the MK-54 torpedo.

The ONR program for the propulsion of undersea weapons has two main thrusts: low-rate energy sources and high-rate energy sources. The low-rate energy sources would be used in unmanned underwater vehicles (UUVs) targets, small delivery vehicles, and other low-speed vehicles, while the high-rate energy sources would be used in high-speed weapons. Energy sources that can be switched from a low rate to a high rate would be applicable in weapons that have a low-speed search mode and high-speed attack.

The conventional low-rate energy sources are electrochemical devices such as rechargeable batteries and fuel cells. ONR activities are concentrated on high-energy-density electrochemical systems, including rechargeable lithium batteries and the new semifuel cells. The semifuel cell is intermediate between a battery and a fuel cell. It has aluminum (or magnesium) anodes that are consumed during operation and an oxidant that interacts with the catalytic cathode.

Research is also being conducted using thermal units to provide low-rate energy sources. The thermal conversion activities include the development of a small, closed-cycle Stirling engine coupled to a lithium-sulfur hexaflouride thermal-energy source. A novel wick combustor is being developed for this unit using capillarity to distribute the liquid metal.

High-rate energy sources are being evaluated for potential use in torpedoes and in countermeasure applications. There are two main ONR activities in this field, HYDROX, a hydrogen and oxygen producer and combustor, and an aluminum-water vortex combustor for a water ramjet.

The HYDROX energy system produces gaseous oxygen from liquid lithium perchlorate and hydrogen from the reaction of water and a lithium-aluminum alloy. The gaseous hydrogen and oxygen produced are burned in a combustion chamber to produce steam for a closed Rankine-cycle system. The same gas source could provide the hydrogen and oxygen for a fuel cell. The gases could also be used in a combined system utilizing a low-power unit for low-speed search and a high-power unit for high-speed operations. The innovative wick system to distribute liquid metal is being developed for use in the SCEPS (lithium-sulfur hexafluoride) upgrade.

A novel vortex combustor is being developed for the water ramjet that would propel the high-speed supercavitating vehicle. Aluminum particles are burned in a vortex arrangement in a reaction with water. This unit, although potentially useful as a source of high-density energy for the supercavitating ramjet, could be used in other applications. The production of large volumes of gaseous hydrogen from the aluminum-water reaction could, perhaps, be utilized to increase the energy density.

The high-rate-wick Stirling engine can be employed in torpedoes and manned undersea vehicles and/or UUVs to enhance range, speed, and endurance. The HYDROX system could be used in high-rate, low-rate, or hybrid modes to enable smaller vehicles or superior performance. The aluminum-seawater vortex device could provide very high speed in special applications. These innovative approaches are good examples of revolutionary technology from ONR programs.

Other propulsion S&T efforts include those on electrochemical energy sources, including fuel cells at the Naval Undersea Warfare Center (NUWC), Naval Surface Warfare Center, Carderock Division (NSWC/CD), and several small academic and industrial contractors. The electrochemical area is the largest component of the undersea weapons 6.1 budget ($2 million in FY99). Another effort is that on underwater propellants at the Naval Surface Warfare Center, Indian Head (NSWC/IH).

Finding: The program on propulsion at the Applied Research Laboratory, Pennsylvania State University (ARL/PSU) is exemplary and offers technologies for both weapons and vehicles that could be used in future systems. Closed-cycle engines are among the increasingly attractive options as the importance of stealth and endurance increases.

Guidance and Control

The operational effectiveness of modern torpedoes has eroded in the face of countermeasures, reflected clutter (or reverberation) in shallow water, and the diminishing strength of acoustic targets. The core issue is the ability of the torpedo's guidance and control system to separate an actual target signature from clutter, whether generated by countermeasures or as a reflection from the environment. During the Cold War, quieting and countermeasures were the major challenges. Now, quiet and small diesel-electric submarines and high ambient noise levels in the littoral shallow waters result in severe problems for target acquisition and homing, including the following:

- Low probability of detection (P_d) owing to high reverberation, small target size, and target aspect dependency in low-Doppler scenarios;
- Severe downward-refracting sound velocity profiles, resulting in a loss of detection opportunities; and
- Low probability of classification owing to the density of environmental false targets (primarily bottom and surface returns).

In addition, there are also two types of problems inherited from the Cold War:

- Prevention or loss of track by multiple jammers; and
- False homing on multiple mobile decoys.

The technology under development for handling weak and false targets in each of these areas focuses on the use of sophisticated waveforms, enhanced processing, and improved sensors. For its undersea weapons S&T program, ONR is applying its limited resources through its programs in broadband processing, advanced guidance and control, and simulation and testing. Enhanced processing and waveforms have the potential to significantly improve performance in a complicated signal environment. Further, the Torpedo Master Plan allows for improvements to be inserted into the inventory through preplanned product improvement. Advances in processors and software technology are the key enablers that allow "smart" behavior in a complicated environment; current limitations are directly attributable to the use of the cumbersome logic necessitated by obsolete processors. These capabilities—when demonstrated in the existing 6.3 advanced technology demonstration (ATD) programs—should improve single-weapon performance: if a detection opportunity occurs (i.e., if there is a target return in the field of view of the weapon sensor), the probability of acquisition (P_{acq}) will increase. The transition path for the broadband processing and intelligent control technologies exists, and near-term improvements are expected.

There will, however, still be significant limitations on what a single weapon can achieve once launched and on its own. The sensor performance of torpedoes is limited by the bandwidth of current transducers, noise sources, and a limited receiver aperture. The committee was unable to determine how critical the additional bandwidth is to effective broadband processing.

Finding: It is not obvious to the committee that the programs on guidance and control at ARL/PSU and NUWC are succeeding at coping with progressively quieter targets and evolving countermeasures. The careful operations and systems analysis needed to critically assess operational performance in matters of target detection, identification, and homing seems to be missing.

Weapon-Platform Connectivity

The most promising approach yet to be investigated to the sensor limitations of a restricted aperture may be the proposed use of platform sensors and processing linked to torpedoes either through fiber-optic cables or acoustic communications. Eventually this approach would extend to offboard sensors, and it holds significant promise for dealing with a wide variety of complicating factors. It could take full advantage of submarine and tactical surveillance sensors, it is a natural step toward cooperating torpedoes, and it fits well into the future vision of controlling the littorals by means of manned and unmanned platforms.

The anticipated replacement of wire guidance links with fiber-optic links, although apparently driven by other factors, would offer an opportunity to make use of the high inherent bandwidth for high-data-rate, two-way communication between the submarine and the weapon, although it is not currently planned to take advantage of this opportunity. There are several potential benefits to be obtained from weapon-platform connectivity, which entails the sharing of platform information with the weapon.

First, owing to the torpedo sonar's limited aperture, its beams are wide and often cannot resolve targets and countermeasures. Using the sensor suite on the platform, contact information could be downloaded to the weapon to improve its acquisition and homing performance. The information from the submarine high-frequency sonars (chin, sail arrays) is in the same frequency band as that from the weapon but is obtained with an order-of-magnitude improvement in spatial resolution. Different aspects presented to the torpedo and the submarine could also facilitate the separation of targets from clutter. The self-noise of the submarine sensor would probably be less than that of the torpedo.

The second potential benefit would be the sharing of platform computational resources with the weapon. Even with modern processing boards, the weapon throughput is limited owing to the small space available. Platform processors could be used to augment the weapon processing as long as the link is sound (e.g., additional weapon beams could be processed with computationally intensive algorithms, and prediction models involving detailed oceanography could be run in real time). If the link is disrupted, the weapon would retain its autonomous capability.

A third potential benefit would be display of the weapon's tactical view on the platform's displays. While the platform has good global information, its local information is not as good as that of the weapon. If the weapon tracks and confidences are shown on the sonar displays, comparisons can be made between the two sensors and differences resolved. In addition, platform interaction with the weapon can be enabled for undersea, surface, and air.

The fourth potential benefit of weapon-platform connectivity would be the enabling of a rapid response when evidence of a threat is seen. In the case of a submarine platform, the weapon could be launched well outside its autonomous acquisition range and with a poor fire control solution. Once it was far enough away to prevent a threat counterfire weapon from seeing ownership, the weapon could turn and transmit into the region of the target. The threat would be detected bistatically by means of the platform high-frequency sensors, using the weapon as a mobile active source. In this way the weapon could move to high speed much sooner than would normally be possible in shallow-water environments. This would minimize the threat's ability to react. In the bistatic mode, the weapon could be transmitting low-probability-of-intercept pulses to give a covert active sonar capability without transmitting from the submarine. In addition, the weapon and high-frequency submarine sonar outputs could be processed coherently to enhance passive detection and to perform target localization by estimating the Doppler components and the time difference of arrival.

Weapon-Weapon Connectivity

The high-bandwidth connectivity between weapon and platform enables another important opportunity. The platform can now act as a node to connect two weapons in the water. The spectral response (or beam aspect, for a monostatic system) of the target gives the best opportunity for detection. The opportunity to detect targets independent of aspect would increase considerably if two (or more) communicating weapons could be utilized. With broadband sensors and processing implemented on each weapon, an attack could be prosecuted wherein each weapon operates monostatically in its own band and bistatically in the band of the coordinating weapon. In this way, search geometries could be devised and would regain much of the robustness provided by the receipt of spectral returns. In addition, with fiber-optic connectivity, the weapons could share raw data, leading to large baseline coherent processing between them and three-dimensional imaging possibilities. This would be especially helpful in multiple countermeasures scenarios and in classifying bottom returns. The intelligent control architecture necessary to manage this type of information and decision making is now available.

Similarly, the depth dependence of the weapons could be dealt with. Current weapons are launched at the best depth for coverage of the water column. However, in shallow water with downward refraction there are large areas that the weapon is not able to interrogate. In addition, frequency-selective fading is best handled by depth changes. Communicating weapons could coordinate their search strategies in the depth plane.

Long-term Implications

The preceding discussion illustrates the need for a system architectural context for undersea weapons guidance and control that encompasses the sensing and processing capabilities of weapons, the platform, and—eventually—offboard sources and or receivers.

Such a program would be tasked with providing an analytic (and simulation-based) foundation to allow development across program boundaries. While there is cooperation between the ONR weapons group and its ocean monitoring group, increased integration will be critical as weapons guidance and control become increasingly linked to a distributed monitoring environment.

Finally, the committee recognizes that advances in autonomous control, precision navigation systems, the Global Positioning System, and underwater communications are likely to significantly change the character of undersea weapons. The future is likely to include neutrally buoyant, long-endurance torpedoes or UUVs. These weapons will be able to launch and loiter quietly. They will work cooperatively with each other and other deployed systems. In doing so, they will extend the reach and security of manned platforms. The committee expects that such weapons, together with air-deployed direct-attack weapons, will be the littoral weapons of choice in the future.

Torpedo Stealth

The committee believes improving the quieting of existing heavyweight torpedoes is of paramount importance. An adversary's use of countermeasures is very dependent on the detectability and classification of incoming weapons. Improved weapon stealth and stealthy launch shorten the time available for an adversary to detect and identify an attack and then deploy effective countermeasures, and they improve the ability of the weapon to acquire a target by reducing self-induced noise.

The ONR efforts in quieting technology for current and future torpedoes supports the PMS 404 (undersea weapons program office) multiyear Stealth Torpedo Enhancement Program (STEP). The objective of the ONR torpedo stealth initiative is to reduce the acoustic and nonacoustic signatures of undersea weapons to improve stealth and reduce sensor self-noise. The technology program appears to be appropriately focused in four primary areas:

1. The quieting of machinery-induced noise by use of advanced vibration damping/isolation methods;
2. The development of quieter propulsors and the quieting of open-cycle exhaust;
3. The development of homing guidance capability using covert passive or low-probability-of-intercept homing sensor technology; and
4. Overall vehicle design, including smaller, lighter, more efficient warheads and propulsion sys-

tems that would reduce negative buoyancy and allow slower run-out speeds, thereby reducing hydrodynamic and propulsion noise.

The stealth technology program has two parts. The first deals with understanding the sources and mechanisms of noise generation and quantifying them, along with improving the models and simulations for predicting the effects of mitigation methods and design changes. The second part focuses on technologies for reducing or controlling noise, such as hybrid active and passive isolation in machinery mounts, and on innovative acoustic design. A long-term objective of this work is to be prepared to support a stealthy torpedo enhancement phase (STEP 2), anticipated to begin in FY08, as shown in the Navy master plan schedule (see Appendix A). It is expected that the capabilities that are developed will also apply to long-endurance, low-speed stealthy vehicles.

In the near term, the stealth technology program is aimed at reducing the signature of the MK-48 heavyweight torpedo and the MK-54 lightweight hybrid torpedo, which uses the old MK-46 lightweight torpedo propulsion section.

One solution proposed by NUWC for the engine noise/hull-coupling problem for both the MK-48 and the MK-54 open-cycle engines is to develop mufflers. Data shown the committee support the view that mufflers will offer some limited levels of quieting for an open-cycle engine; however, the cost-effectiveness of their incorporation into the current torpedo inventory should be assessed.

The heavyweight torpedo has significant hydroacoustic noise related to the high speeds at which it must travel to generate enough body lift to overcome its negative buoyancy. ONR's program does not appear to include a significant effort to reduce that noise.

One approach to speed reduction and consequent mitigation of radiated hydroacoustic noise would be to provide the MK 48 with pop-out lift surfaces (wings) so that all lift does not come from body lift, which can be developed only at high speeds. The committee believes that the ONR undersea weapons S&T program should explore a variety of hydrodynamic solutions to the hydroacoustic noise problem.

The ONR torpedo stealth technology work planned and under way appears to offer an encouraging long-term payoff: much quieter next-generation weapons. As mentioned above in assessing the propulsion technology, good progress is being made, particularly at ARL/PSU, in innovative, closed-cycle power sources and in quieter propulsors that will allow slow run-out speeds with high speed after the weapon acquires the target and begins homing. These systems will be inherently quiet as well as compact. However, since the MK-50 SCEPS turnaround cost is cited as one reason for the MK 54's use of a noisy MK-46 open-cycle engine, care must be taken to consider the cost of ownership when developing these systems.

The reductions in weight and volume resulting from improved propulsion and warheads, together with lighter composite hulls, should support the ability to operate at lower speeds with sufficient lift to reduce self-induced sensor noise prior to target acquisition. Covert target acquisition and homing guidance performance should also be enhanced by reductions in radiated noise.

On balance, while the ONR torpedo stealth technology program appears to be well integrated with other technology areas, particularly in the longer term, quantitative analysis is needed to determine the payoff of the stealth effort. The committee notes that unless significant and extensive retrofits are allowed to existing inventory, quieting gains on those weapons will be marginal. It might be more cost-effective to forward-fit these improvements into new-generation weapons than to try to retrofit them.

Finding: Upgrades intended to quiet the MK-48 and MK-54 torpedoes (mainly by NUWC) were not persuasively presented to the committee. The open-cycle engine, buoyancy disadvantages, hydroacoustic noise, and other characteristics make the upgrades questionable in light of the evolving stealth and countermeasure capabilities of potential enemy targets. No systems analyses of predicted program success or time scales for acquisition were presented to the committee.

Torpedo Defense

Although the U.S. Navy devotes a large and continuing effort to defeating missile attacks against surface ships, the effort devoted to the protection of U.S. submarines and surface ships against undersea weapon attack is relatively miniscule and far too small relative to the effects of a single hit. For FY99, for example, the ratio of funding for torpedo defense to that for shipbuilding was about 0.12 percent. The ONR undersea weapons S&T program does have a component for the development of technology for own-ship defense. The committee finds these efforts to have varying degrees of merit and is concerned that there has apparently been no operations and systems analysis of the problem that could be used to guide the directions of developments.

Through the years a number of concepts for defense against torpedo attack have been postulated, including the following:

- Noisemakers;
- Mobile decoys;
- Supercavitating pellets (undersea weapons analog of the Close-in Weapon System);
- Antitorpedo torpedo (e.g., 6.25-in. supercavitating weapons); and
- Adaptations of the current MK-46 torpedo to antitorpedo torpedo operations.

Individually, each of these technologies, if properly integrated into a coherent defense system, might provide considerable added capability. However, the committee heard no evidence that a coherent view of the problem is used in deciding the allocation of budgetary resources.

Surface ship defense against missile attack is driven by a coherent concept of operations (CONOPS). There is no equivalent doctrine for torpedo defense. In the missile defense CONOPS, surveillance radars on surface ships and aircraft detect the presence of incoming hostile missiles. When the detection and target identification process has been completed and when tracks of the incoming missile have been formed, a long-range weapon is assigned to engage the target. In the case of surface ship defense, weapon guidance is semiactive and a fire control radar is tasked with providing weapon guidance. Decoys may be launched simultaneously in an attempt to spoof the guidance of the incoming missile. If long- and mid-range weapons fail to engage and destroy the incoming missile, close-in weapons are employed.

The performance requirements for the analogs of this process for torpedo defense have not been enunciated. What sensors will be used to detect and track the incoming torpedo? With what accuracy can its position and velocity be defined? Are the available sensors capable of providing the requisite accuracy? At what ranges should the incoming torpedo be engaged? What is the likely firing doctrine? How will defensive weapons be guided to their targets? What will be the role of countermeasures (noisemakers, mobile decoys, and so on), and what would be their attributes?

Although it is not ONR's responsibility to come up with a full-fledged torpedo defense doctrine, such a doctrine is needed to judge the value of various efforts related to torpedo defense within an overall system context that has been established in enough detail to identify general performance thresholds.

Decoys, Countermeasures, and Counter-Countermeasures

An asymmetric threat includes weapons of precision attack but limited firepower. Torpedoes, smart mines, and diesel submarines are the undersea threats. Cruise missiles are the surface threats. These weapons are characterized by their ability to precisely inflict damage to a capital ship or in a localized region. Their vulnerability is the targeting and control systems used to accomplish their mission. For decades now, the United States has developed countermeasure systems to disrupt the ability of smart weapons to damage their targets. These countermeasures range from relatively unsophisticated offboard jammers that work by saturating the sensor system of individual platforms, to more sophisticated onboard systems that disrupt the weapon control systems. While protection for air platforms has evolved to the point that there is an effective synergy between onboard and offboard countermeasures, protection against undersea threats is not as well developed. This is so, despite the fact that naval forces must be able to counter these threats to be effective in the modern world.

The Future Naval Capability (FNC) thrust in platform self-protection is directed to this requirement. It has as its highest priority platform protection from torpedoes and mines. Clearly, Navy leadership attaches a high priority to such protection, and this should influence the ways in which funding is allocated.

Finding: A number of plausible approaches to defending against torpedoes were broadly outlined to the committee, including noisemakers, decoys, supercavitating pellets, and antitorpedo torpedoes. Individually these might be of value, but maximum benefit will be achieved only if they are integrated properly into a plausible, coherent defense architecture system.

Supercavitating Weapons

ONR's S&T efforts to develop supercavitating torpedoes with speeds of up to 200 knots appear to have been spurred by Russian work, which, in turn, was allegedly motivated by earlier U.S. Navy work on supercavitating weapons that were launched from aircraft or ships.

The current ONR program has three stated objectives:

1. To understand the physics of supercavitating flow;
2. To develop vehicle control and guidance methodology for maneuvering and homing of high-speed vehicles; and
3. To design and build a testbed that can be used to evaluate candidate control and homing concepts.

This is a program that can energize young engineers and scientists. It offers substantial technical and system challenges in launching, hydrodynamics, acoustics, guidance and control, and propulsion, to name a few. While technical progress (analytical and experimental) has been reported, many more difficult problems remain. The program is being used as a rallying point for other desirable activities such as the development of up-to-date design optimization tools and computational fluid dynamic techniques. The value of the technology and tool synthesis planned around this S&T thrust should not be underestimated. The ability of the thrust to attract talented researchers and engineers and to develop new tools is serving the overall S&T program well. For instance, the study of gas cavity/water/vehicle interaction has potential value for other uses in less ambitious areas, such as vehicle control by injecting combustor exhaust bleed gas into the boundary layer.

The committee believes that the testbed approach being planned is appropriate for exploring the phenomenology and for helping in the development of reliable predictive modeling tools.

However, the committee notes that the road map for this combination of activities is ambitious and might be viewed as out of balance with the payoffs the activities offer in the overall undersea S&T program. Although ONR expressed confidence that the technology challenges will ultimately be overcome, no analysis was presented to the committee that makes a convincing argument for the operational utility of the close-in, fast-reaction weapon. A more likely first application might be as a terminal stage for a stealthy UUV first stage, where the vehicle can be pre-aimed at the target by the UUV and inertially guided to achieve a kill. Care must be taken not to oversell the concept on the basis of incomplete application analysis. The committee believes that there should be a careful evaluation of the operational utility of the concepts this technology could enable if it is successful.

Finding: The program at ARL/PSU and NUWC to develop a high-speed, supercavitating vehicle is challenging and sufficiently promising to warrant research in (1) the physics of supercavitating flow, (2) vehicle control and guidance methodology, and (3) the design and building of a testbed. There is a need for careful analysis of the operational utility of the concepts this technology could include.

Weapons Design Optimization

The weapons design optimization (WDO) program category appears to be a relatively recent effort, with funding first appearing in FY98 and continuing through the Future Year Defense Program. The committee endorses the overall vision and basic principles of such a process but remains concerned that the actual implementation will not satisfy the need for operations and systems analysis, as is discussed in Chapter 3.

The overall vision, or goal, is to develop a distributed architecture that integrates performance evaluation, life-cycle factors, and design optimization into a single methodology for use by the entire acquisition community, including government and industry. The early efforts are divided between architecture and model development and selected applications. The development efforts focus on collaborative environments for multiple users, optimization techniques, simulation environments for performance evaluation, and physics-based models to faithfully represent hydrodynamics, propulsion, warhead effects, and so forth.

A number of initial applications were cited, including the Common Broadband Advanced Sonar System, the low-endurance, low-frequency active surveillance, and the 6.25-in. weapon and warhead. Few details were presented as to the exact nature of these applications. It is presumed that since the WDO program is in its early stages of development, an end-to-end application of the process is unlikely for some time, but the committee lauds ONR for parallel development and application, provided that feedback is used to improve the process. Furthermore, the committee believes that successful development and application of WDO could help to alleviate a serious concern, namely that the operations and systems analysis was not adequate to support development and evaluation of the weapon system concept.

While WDO could become a valuable tool for fine-tuning parameters to maximize performance, minimize total ownership costs, and so on, it is only a tool and cannot replace creativity or sound judgment. Future weapons must be deployed within the construct of an end-to-end system, including distributed sensors, unmanned vehicles, network processing, and multimode communications. Therefore, giving WDO sufficient scope, yet enough flexibility to accommodate a range of system concepts, is a significant challenge that will probably not be achieved in the context of any near- to mid-term weapon. As an adjunct to this effort, a higher level oversight process could be created to ensure that the weapon concept being optimized is indeed being optimized within the appropriate context.

The committee was impressed with the process used in the Navy's Advanced, Rapid, Commercial-off-the-shelf Insertion (ARCI) program[1] to develop and upgrade submarine sonar, particularly those aspects of the program that involve peer review and participation, extensive use of at-sea data, validation of candidate algorithms, use of commercial products, and so on. WDO appears to offer a framework within which many of the same principles as those of the ARCI process can be followed. The collaborative environment can allow for peer review and participation and is therefore a good first step. The difficult next step, employing at-sea data, is vital to ensure the validity of the models on which WDO is built but was not apparent in information presented to the committee. The committee believes that ONR should carefully examine, and revise if necessary, the WDO process to ensure that at-sea data, both existing and to-be-collected, are used to the fullest extent in validating the individual models as well as the end product in its entirety.

Overall, the committee believes that ONR should continue to develop WDO but should incorporate it into a larger context that will truly optimize future weapons for future engagements against future adversaries and targets as discussed in Chapter 3 of this report.

Finding: Weapons design optimization, which appears to be a relatively recently identified effort, while useful still does not satisfy the need for operations and systems analysis called for at several points in this report.

TECHNOLOGY ISSUES

Maturity of the Key Technology Areas and Associated Challenges

As presented to the committee, the ONR undersea weapons S&T program is largely devoted to developing incremental improvements that can be inserted into either or both the MK-48 and MK-50 torpedoes. In the case of the MK 50, the plan is to retrofit with MK-46 open-cycle engines and replace the directed-energy warhead with a bulk charge. The MK 50 has a SCEPS engine that will be replaced with an MK-46 engine in the MK 54. The MK-54 hybrid, now in low rate initial production and developmental tests, will also have commercial-off-the-shelf (COTS) components and MK-48 processing software.

The MK-48 program is evolutionary in scope and philosophy. No capability will be developed if it cannot be brought to fruition in time to meet a specific insertion opportunity on the MK-48 road map. An incremental program such as this will certainly improve the performance of the MK 48, but it is unlikely to result in a new weapon with important new capabilities for antisubmarine warfare.

The committee understands the philosophy and budget environment that have driven the undersea weapons S&T program to adopt this evolutionary approach. It hopes that in future years the existing program will be rebalanced and that a larger share of it will be devoted to efforts that are more revolutionary than are current efforts.

Although ONR undersea weapons S&T is tightly coupled to Program Manager, Naval Sea Systems Command (PMS) 404's road map and so is of necessity evolutionary in nature, the committee observes

[1]See Appendix B for further details.

that some of its components could bring about changes that might be characterized as relatively revolutionary. Specifically, the effort on propulsion energetics that is being pursued at ARL/PSU will, if successful and introduced into operational torpedoes, permit a decrease in the length of a torpedo that is assigned to propulsion. An analogous change in space required could come from explosives S&T. This would allow for significant changes in torpedo design. The committee believes that ONR should continue to support revolutionary research that would have as much potential as the energetics program.[2]

There are a number of attractive challenges in the undersea weapons area. In particular, four technological concepts that would appear to be quite useful are not being integrated into weapon solutions: distributed sensor fields, with their attendant integration and display capability; unmanned undersea vehicles; air- and ship-launched direct-attack weapons; and remotely controlled fixed and mobile mines. The development of a distributed sensor field and of the integration necessary to transform individual sensor data into a theaterwide picture is clearly germane to countering an enemy threat in the littorals. Nevertheless, although some elements of technology that might be related to these capabilities are being pursued, they are not part of a clearly defined architecture. For instance, the acoustic communications work appears to be focused on submarine communications and is not really envisioned as the glue needed to make a distributed sensors system useful for conducting antisubmarine warfare in the littorals.

Similarly, the unmanned undersea vehicle work is focused on mine countermeasures and is not envisioned as the detect-and-kill platform of tomorrow. Given that the negative buoyancy of current torpedoes is driving the need for high speed and that their range is limited by their tethers, the need to quiet the MK-48 torpedo could be eliminated by replacing it with slow-moving UUVs loaded with sensors and explosives, as needed.

While there is at least some ONR effort on the technology of communications and unmanned undersea vehicles, there is no work in the undersea weapons program on air-launched direct-attack weapons and offensive mines. Clearly, there would be advantages to an air-launched, short-time-of-flight, direct-attack weapon, supported by a distributed field of short-range sensors, for which a detection often amounts to a localization. Offensive minefields, capable of protecting U.S. projection forces from attack by enemy submarines, would also be relevant in the littorals.

These four are hardly an exhaustive list of weapon concepts that might improve ASW capabilities. There appears to be little in the ONR program that is geared to developing either revolutionary concepts or the technology that would be needed to implement and exploit such concepts.

In short, weapons upgrades are mature, but there are a number of attractive future challenges and concepts that cry out for exploratory investigation.

Interaction with Related Technology Areas

In general, the committee found that the ONR undersea weapons S&T program had little interaction with related technology areas. What interaction there was seemed to be focused along program lines (i.e., in the lightweight or heavyweight torpedo programs) rather than technology lines. This paucity of technology interaction seemed to extend to common issues between torpedoes and UUVs as well as to

[2]Properly speaking, the term "revolutionary" should be used in the sense of changes in capabilities, doctrine, and techniques that will significantly change the execution of antisubmarine warfare. New weapons and weapon delivery platforms will be needed for the implementation of revolutionary concepts.

technology areas outside the undersea weapons arena. Each project appears to stand on its own, and there is little evidence of a concerted effort to take advantage of technologies in related areas. The committee thought this was symptomatic of the narrowness of the evolutionary approach to S&T in the weapons area and suggested that there should be much broader participation by the technical community as well as the application of innovative thinking. The committee is aware that problems related to network-centric warfare are being addressed by other components of ONR, such as those involved in the littoral antisubmarine warfare Future Naval Capability.

For example, there should be greater synergy between the ONR UUV program and the torpedo program. At the very least, communications protocols, navigation, energy transfer, and control language should be standardized across ONR UUV and torpedo programs.

Another example is the apparent lack, at the time of this writing, of obvious program ties between work in the undersea weapons S&T program and the work on undersea weapons supported by the Defense Advanced Research Projects Agency (DARPA).[3] By DARPA standards, its efforts in undersea weapons S&T are rather small. Nevertheless, the committee found the DARPA efforts to be of a high quality and to constitute an important independent approach to the problems in the field. Although the managers of ONR's undersea weapons S&T program are well versed in the DARPA approach, there was no apparent effort to leverage that approach or to incorporate it into ONR's own program.

The DARPA effort is based on the concept that success in future undersea warfare (particularly in littoral regions) will involve the use of networked sensors and weapons. The committee is in general agreement with this concept and believes that it should have more influence on the ONR undersea weapons S&T program, as suggested above in the "Long-term Implications" section under "Guidance and Control." Quantitative operations analysis applied in this area, as in others, could assist in comparing the payoffs of different systems and CONOPS.

In summary, interactions with related technology areas need strengthening. The committee believes that establishing more useful interactions between the management of the ONR undersea weapons S&T program, the DARPA undersea warfare program, and the ONR antisubmarine warfare S&T program would result in a more modern and forward-looking approach to undersea weapons research.

Missing or Inadequately Addressed Undersea Weapons S&T

The committee believes that ONR's undersea weapons S&T program should be planned and implemented in a broader context than was evident from the ONR presentations. The challenge of littoral undersea warfare requires continued new thinking and concepts of operations that are distinct from past Cold War tactics and strategies. The following discussion of developments not being addressed by the ONR undersea weapons S&T program, for example, necessarily reflects the knowledge and interests of the members of the committee. The need for S&T is a constant, and ONR must maintain a reservoir of S&T from which it can draw for weapons development. As discussed below, the committee believes that S&T needs to go beyond the torpedo in its present or upgraded form.

The advanced deployable surveillance systems in development will evolve to provide multiple-phenomenology sensing and connectivity that can be quickly deployed in any location.[4] Acoustic

[3]The Navy master plan road map (see Appendix A) indicates an insertion of this technology as an MK-54 product improvement in 2001.

[4]At the present time, the advanced deployable system requires a cable termination on a friendly shore.

communication technology, which is being developed under ONR 32 sponsorship, offers the opportunity for relatively high bandwidth connectivity that can be synergistically deployed with or added to the arrays.

The challenge of effectively confronting submarines in the littoral environment is not being addressed on the basis of a complete analysis and understanding of the physics involved. Acoustic methodology has long been the primary means of underwater communication and target identification and location, and the Navy has consistently been a leading performer in this area. The intention of the present ONR undersea weapons S&T program to make full use of the acoustic bandwidth available is endorsed by the committee. At the same time, however, the committee is concerned that the understanding of the inherently complicated and variable acoustics of near-shore waters is incomplete and that the present performance objectives may not suffice to defeat the small, quiet submarine and other potential threats. The problem should be attacked at its most basic level.

Fiber-optics bandwidth is extremely broad, especially in the context of acoustic communications, greatly benefiting the speed and density of transmission. Computational facilities can be located far away from the weapons, and minimal onboard computational power is necessary when broadband paths are available. There are important undersea weapons applications of this technology to be exploited, problems of tethering notwithstanding.

The committee recognizes the advances made in undersea propulsion by ARL/PSU. The automotive industry has also made much progress in the use of internal combustion engines coupled with modern batteries. This technology should be monitored and pursued as it could enable undersea activities at favorable cost. The success of the diesel-electric submarine on the world market suggests that this type of propulsion technology should be of interest for hybrid weapons carrier concepts.

Deployable sensor arrays and communication elements can enable effective littoral undersea weapons capability. Such systems when networked together can provide a capability to fuse distributed-sensor data into a tactical picture, allowing anything that enters the waters containing these arrays to be detected and localized. Accurate air-launched, direct-attack weapons, or shipborne rocket-delivered standoff weapons, could be targeted based on that data. No such standoff or direct-attack weapons concepts were mentioned in the information provided to the committee by the antisubmarine or air warfare communities.[5] S&T efforts aimed at developing such a concept should form part of the future undersea weapons program.

Low-probability-of-intercept bottom transponders seeded among distributed surveillance arrays also can allow UUVs to navigate freely in that area using occasional time-difference-of-arrival fixes from two or more transponders to update the UUV navigation system. The bottom transponders, each equipped with a cheap Global Positioning System/inertial navigation system (INS), would, as they were deployed, store a fix just before entering the water, and the INS would measure only the drift and drop from deployment until coming to rest on the bottom. Each such bottom transponder, knowing its location, could transmit those data upon interrogation, along with the time of arrival and processing time of the interrogation. Using time-difference-of-arrival plus the transmitted positions from two or more of these transponders, a hunter/killer UUV equipped with a low-cost, cruise-missile-quality inertial guidance system having a similar low-probability-of-intercept transponder could communicate with the

[5]The MK-80 bomb hydrostatic fuzing work under ONR could be regarded as directed toward a minimal air direct-attack capability, but it is a depth bomb, not a rocket-propelled weapon. As was done in World War II, analysis is needed to determine the utility of any depth bomb approach.

bottom transponders and fix its position periodically at known times of the day, obviating the need for a very accurate INS capability on board.

It is also feasible to communicate data to and from roving UUVs for targeting and enabling weapon release or recall and to receive data about UUV status and its operations. Bottom transponders could act as two-way store-and-forward communication links by communicating acoustically with sonobuoys that could link via satellite communication or line of sight to receivers at a command and data fusion center aboard ship. In this way, UUVs could be directed to a different search area, overridden, or recalled at any time. Similarly, UUVs could send status and mission data back to the command ship. Presumably the sonobuoys would also be used to receive data from the distributed arrays in locales where fiber-optic links to the command center are either inappropriate or unsustainable.

Finally, the FNC integrated product team approach embarked on by ONR has the potential to provide high-level, near-term guidance, but the operations analysis and systems engineering required to translate that guidance into a coherent program is not in place. The committee believes that the top-level architecture for obtaining related undersea weapons should be articulated and that a single organization should be identified and charged with responsibility for providing operations and systems analysis. This is not a criticism of the ONR undersea weapons S&T program per se but alludes to a broader issue. Members of the committee believe this lack is a serious problem and address it as a major issue in Chapter 3.

It was the consensus of the committee that Navy-sponsored R&D is required to meet all of the needs identified in the foregoing discussion. The key elements of many enabling technologies for new undersea weapons architectures and concepts of operations are already being pursued at some level within ONR and/or DARPA, but better coordination and integration between program elements will be required.

The consensus of the committee is also that more effort will have to be devoted to operational concepts and to operations and systems analysis of the enabling technologies to validate their utility and implications. This effort should involve a broad spectrum of Navy, academic, and industrial expertise and should lead to criteria for, and an architecture for, the main weapons development programs of the future.

INFRASTRUCTURE

U.S. Performer Base[6]

The ONR S&T programs are primarily developed by the program officers and by the Naval Research Laboratory and its international field offices. The committee found it difficult to determine the total number of professionals who work in the general field of undersea weapons S&T. However, based on its general contacts with organizations that work in the area, it was able to estimate that while between 600 and 800 people are involved full- and part-time in all aspects of the undersea weapons S&T program, only about 250 actual work-years were accrued in FY99 (Figure 2.3).

---

[6]The foreign torpedo performer base is commented on in the next section, "Leadership."

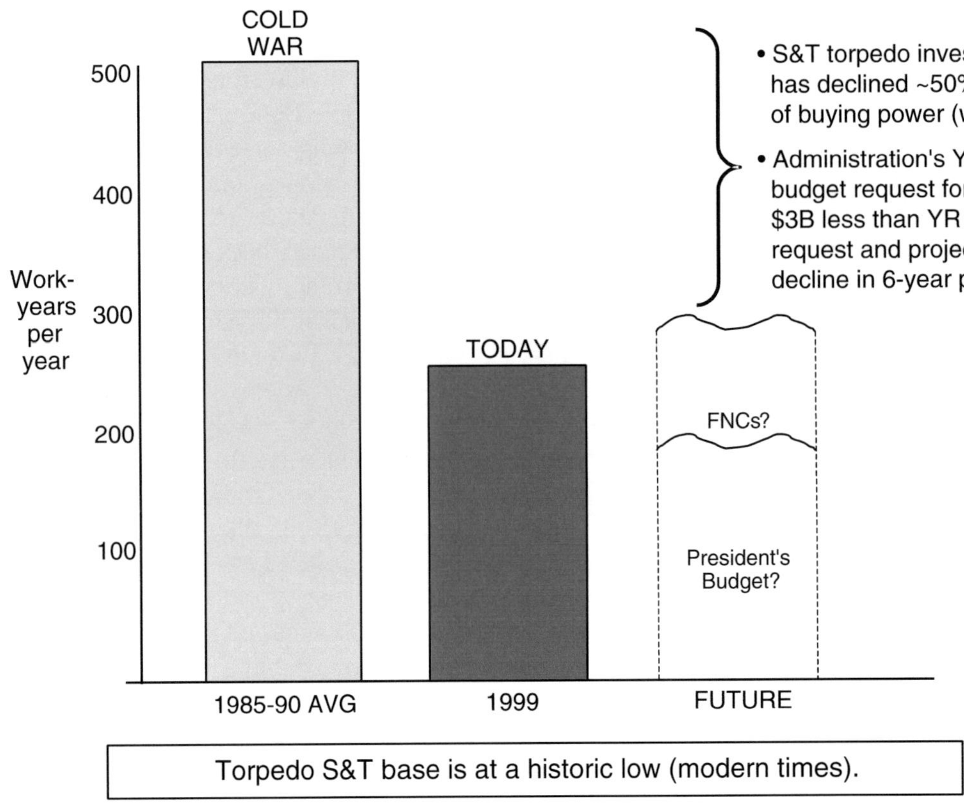

FIGURE 2.3 Torpedo S&T base.
SOURCE: Juergen G. Keil, "Summary: NUWC Weapon Perspectives," Naval Undersea Warfare Center, Newport, R.I., briefing to the committee, November 10, 1999.

Diminishing Industry Participation

The expertise in undersea weapons that exists in industry and universities is barely utilized. Just as government S&T spending has dropped, industry's discretionary funding has also dropped and is subject to internal competition for funds. Generally, industry can afford to spend less than 5 percent of its sales on R&D. When it can see a potential for attractive sales, it is inclined to spend its discretionary money on strategies to help win those sales. The converse is also true, and industry cuts its R&D efforts if there is no sales potential on the horizon. Moreover, the ONR strategy stated in the Navy master plan is to keep exploratory development and technology demonstrations in-house to ensure that the Navy maintains some resident corporate expertise. According to data provided by the sponsor in FY99, less than 10 percent of ONR undersea weapons S&T funds went to industry (including analysis support at federally funded research and development centers and related think tanks), and 94 percent of the funding for university research in underwater weapons went to just two institutions. This situation is not in keeping with the Department of the Navy's aim of providing outreach to the science and technology community.

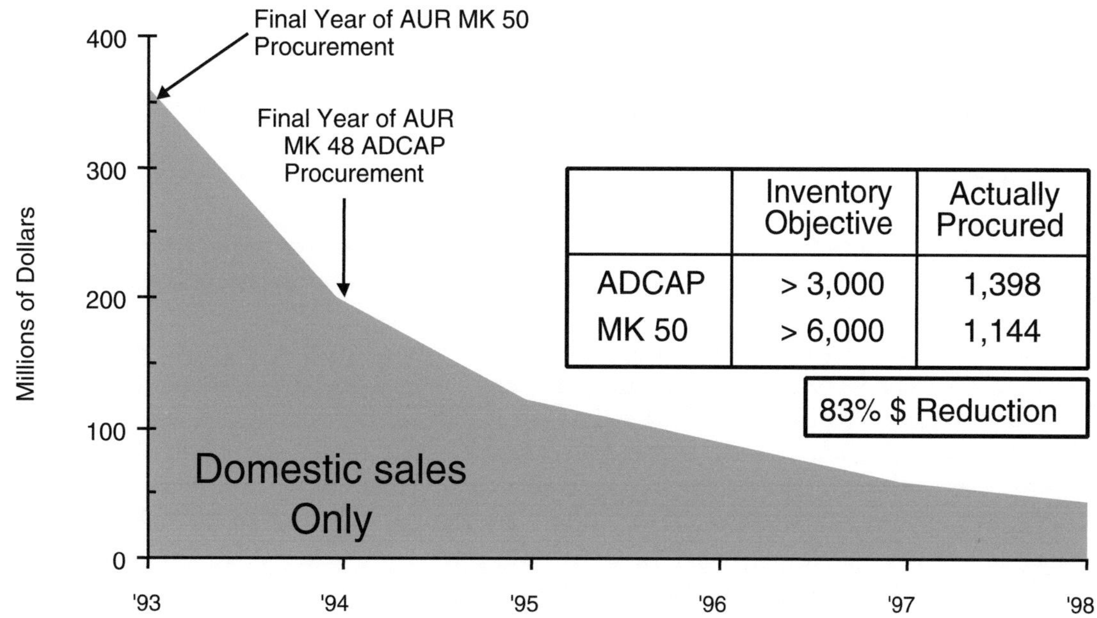

FIGURE 2.4 U.S. Navy new torpedo procurement as indicated by the combined U.S. sales of Northrop Grumman and Raytheon. Acronyms are defined in Appendix D.
SOURCE: John Zittel, "Undersea Weapons S&T," Office of the Chief of Naval Operations, N84T, Washington, D.C., briefing to the committee, October 18, 1999.

Diminishing Base of Technical Expertise

Clearly, talented students, engineers, and scientists, whether in industrial, government, or university laboratories, will not stay in or enter this field if they are not given the resources for meaningful work that will advance the field. Industry will invest its money and its talent in areas that offer a better rate of return. As has been the case in many other areas, the Navy has some fundamental decisions to make if it wants to have a coherent strategy for preserving the expertise for torpedoes. Is it to be a strategy based on industrial expertise, or an arsenal strategy going back to the old torpedo-factory and government-laboratory approach? Section III of the Navy master plan suggests it has chosen the latter.[7]

Whatever the strategy, a situation in which there is no significant procurement of torpedoes virtually ensures that there will not be any substantial industrial investment in related science and technology. Figure 2.4 shows the recent history of the procurement of torpedoes from industry. A significant component of the industrial base is the engineering and scientific expertise at both industry and government laboratories. Past experience indicates that once it is lost, such expertise takes about 10 years to reconstitute, if indeed the people can be attracted. This compares with the cold restart of an existing product line, which took from 3 to 5 years.

It is therefore vital to have an active, robust Navy-sponsored S&T program in place, ready for the

[7]Department of the Navy. 1996. *Undersea Weapons, Vehicles, and Countermeasures: Master Plan,* The Pentagon, Washington, D.C.

time when a new suite of undersea weapons must be acquired. The program must be robust enough to attract young graduates and postgraduate researchers, and it must focus not on today's paradigm but on the capabilities that might be needed 20 years from now.

The committee observed, however, that just the opposite appears to be happening:

- Torpedo S&T, development, and production investment is at its lowest level since before World War II, and S&T funds are not enough to meet current requirements.
- The nation's industrial infrastructure for torpedo development and manufacture, particularly its perishable human expertise, is rapidly atrophying because all new torpedo production has been discontinued.
- Fleet training and stockpile test firing rates are dangerously low, as shown in Figure 2.5.
- Torpedo technology expertise is concentrated at a few government-dominated laboratories (NUWC and ARL/PSU), and competition of any sort has all but been eliminated.

In the past, similar situations led to torpedoes and torpedo shooters that were unable to function in times of conflict. For example, before both World Wars, torpedo R&D and production became the sole responsibility of NUWC/Newport, and that effort was critically underfunded. Insufficient testing of torpedoes and fleet training led to disastrously poor performance that was not remedied until years after the United States had entered the wars.

The committee is concerned that the nation is headed down exactly the same path, with very predictable outcomes. An oversight procedure is needed that will, on a continuing basis, critically examine, assess, and report on the status of fleet torpedo readiness (both inventory and training) as well as on all levels of undersea weapons research, development, testing, evaluation, and production.

In short, the number of participating companies (mainly Northrop Grumman Corporation and Raytheon Systems Company), academic institutions (mainly Pennsylvania State University and the University of Rhode Island), and government laboratories (mainly NUWC/Newport and NSWC/Indian Head) in the United States is small. The number (about 12) of foreign industries and government institutions in torpedo technology and development is larger.

Finding: The committee believes that a truly healthy undersea weapons S&T program should have industry participation, but industry is not now a significant participant/investor in undersea weapons S&T.

Leadership

Basically, the Navy has consolidated its torpedo R&D infrastructure at three locations: (1) the Naval Undersea Warfare Center at Newport, Rhode Island, and Keyport, Washington, (2) the Applied Research Laboratory at Pennsylvania State University, where the Navy's R&D efforts on guidance and control, propulsion, and hydrodynamics are centered, and (3) the Naval Surface Warfare Center facility at Indian Head, Maryland, where work related to warheads and energetic materials is concentrated.

Industry is only engaged in support for these activities. The decision by the Department of Defense and the Navy to shut down all new torpedo production in this country has resulted in the loss of human expertise in this field, as indicated by data in Box 2.1.

While not directly related to the S&T program per se, the torpedo inventory and acquisition strategy seem out of sync. The funding data provided to the committee suggest that fleet requirements cannot be supported, particularly if planned sales to foreign military take place. It is not clear whether the funding

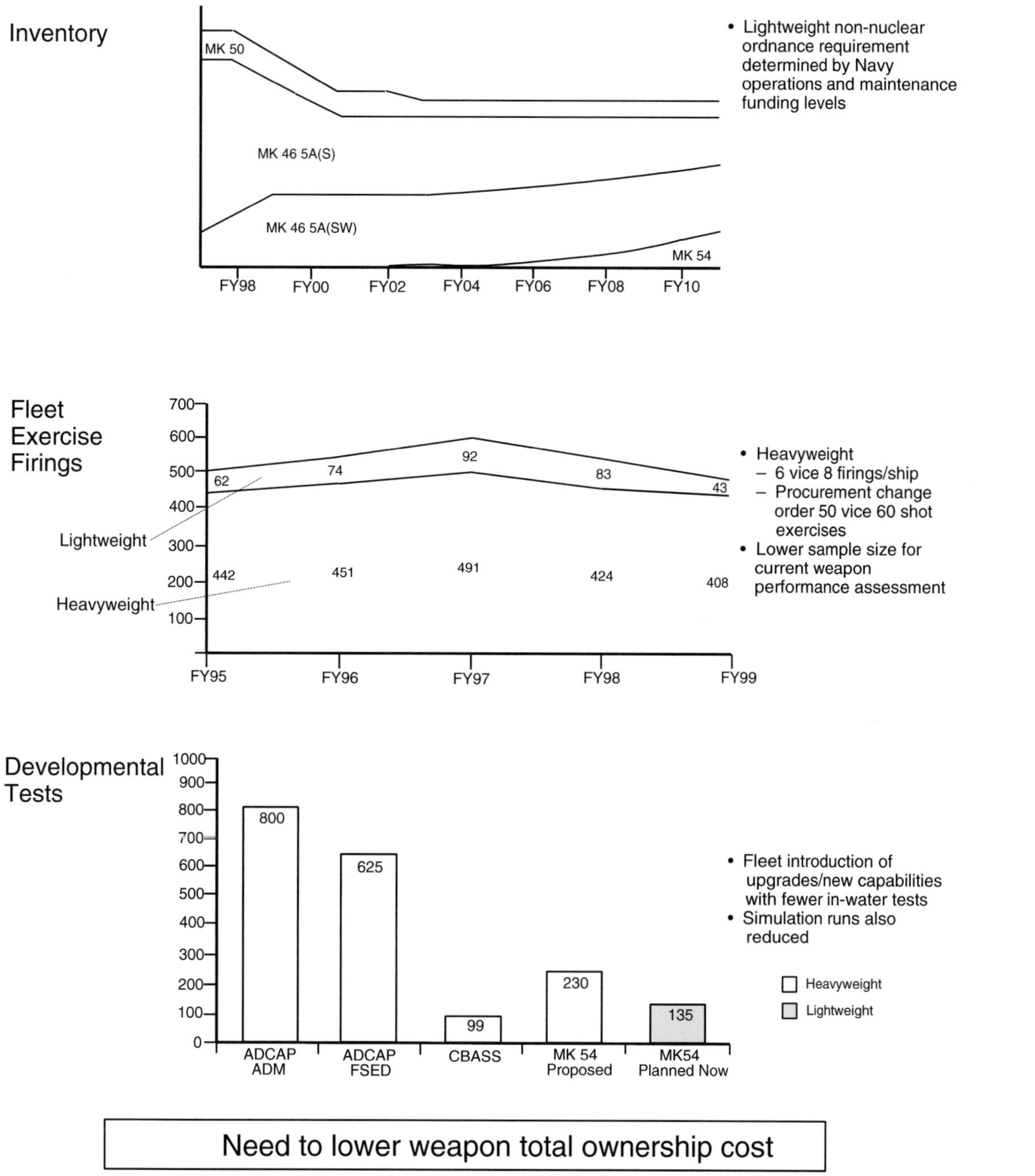

FIGURE 2.5 Impacts of a budget-constrained environment on torpedo inventory and testing.
SOURCE: Juergen G. Keil, "Summary: NUWC Weapon Perspectives," Naval Undersea Warfare Center Division, Newport, R.I., briefing to the committee, November 10, 1999. ADCAP, advanced capability; ADM, advanced development model; FSED, full-scale engineering development; CBASS, Common Broadband Advanced Sonar System.

Box 2.1 Change in Infrastructure and Performer Base Reflecting Change in Market Size

Northrop Grumman Corporation
- Significant reduction in workforce (88 percent)
  — Barely maintaining critical skills
- Facility reduction (97 percent)

Raytheon Systems Company
- Hughes Aircraft Company acquired Alliant Technology and consolidated; then Raytheon acquired Hughes
  — Shutdown of Hami, Mississippi, torpedo facility
- Reduction in workforce (greater than 47 percent)
- Business area consolidated at Portsmouth, Rhode Island
- Facilities in Poulsbo and Mukilteo, Washington, relocated to Naval Undersea Warfare Center, Keyport, Washington

SOURCE: John Zittel, "Undersea Weapons S&T," Office of the Chief of Naval Operations, N84T, Washington, D.C., briefing to the committee, October 18, 1999.

requests also reflect funding for ships in refit or overhaul, but in any case they appear inadequate to provide for a surge capability in time of crisis. Presumably, the policy of selling torpedoes to foreign military was an attempt to preserve the industrial base as much as possible. Yet if there is no procurement to replace that inventory loss, the intent of the policy is defeated.

There is a significant connection between this issue and the undersea weapons S&T program. Just as government spending on science and technology has shrunk, industry discretionary spending has also diminished. Several foreign nations are quite active in the production of a variety of torpedoes that compare well with U.S. torpedoes. Box 2.2 notes a recent assessment by the Office of the Chief of Naval Operations for Antisubmarine Warfare Requirements (N84). The committee believes that the U.S. Navy must make a stronger effort to provide leadership in undersea weapons development so as to match the activity and capability of other nations.

Knowledge Base

The undersea weapons S&T community draws on many areas of professional expertise, among them the following:

- Hydrodynamics;
- Structural acoustics;
- Acoustic sensors and signal processing;
- Underwater acoustic propagation and environmental effects;
- Propellant and energetic material chemistry;
- Energy conversion and propulsion engineering;
- Computer design and software development; and
- Systems and operational analysis.

Box 2.2 Current U.S. versus Foreign Torpedo Technology, with Indication of European Efforts Ahead of U.S. Undersea Warfare Weaponry

United States

- Single conductor duplex wire
- Digital
- No offboard sensor
- Quieting technology
- Broadband array
- Wake homing invented here
- Exercise cost, $10,000 to $30,000
- Thermal propulsion
- Warhead large, high-performance PBXN
- Salvo 2
- Procurement cost, $0.35 million to $1.2 million
- No current production

Foreign

- Fiber-optic cable
- Digital neuro-net processors (artificial intelligence
- Offboard sensor (unmanned underwater vehicle)
- Quiet because of low-speed capability
- Broadband array and knowledge of littoral
- Wake homing in use
- Exercise cost, $8,000 to $250,000
- Electric, thermal, and rocket propulsion
- Warhead medium, 11BX
- Salvo 2, 6, 12
- Procurement cost
  — $1.1 million to $2.3 million (Europe)
  — $0.6 million to $3.5 million (Russia)
- Current production of advanced lightweight and heavyweight torpedoes

SOURCE: John Zittel, "Undersea Weapons S&T," Office of the Chief of Naval Operations, N84T, Washington, D.C., briefing to the committee, October 18, 1999.

As pointed out above, the total number of professionals in the undersea weapons S&T community may be 600 to 800 people. Even if turnover is as high as 10 percent per year, the total number of accessions needed to support current levels of effort should not be more than 60 to 80 people a year. To date, the few institutions engaged in undersea weapons S&T have been able to maintain their performer base with not too much difficulty.

However, the committee believes that if the nation decided to increase its level of effort, it might be very difficult to quickly augment the undersea weapons S&T community. The core disciplines needed for undersea weapons, as well as for other activities, are in place at many universities. Indeed, ARL/PSU offers doctoral-level training in fields that are supportive of, or related to, undersea weapons S&T. Every year the number of graduates of the ARL/PSU program exceeds the number of hires by institutions working in undersea weapons S&T, and so some of these graduates migrate into other fields. Because of overall funding limitations in the field, relatively few new people are entering the field, and the knowledge base is aging.

By contrast, in the specialized disciplines that are unique to undersea weapons S&T, there are relatively few opportunities for graduate and postdoctoral education. These disciplines are not attractive to students because they are thought to be too highly specialized, so they cannot compete with other opportunities. Only oceanographic institutions and ONR-supported laboratories, plus a few university programs, provide the knowledge base necessary for the continuation of a national undersea weapons S&T program. The knowledge-base pipeline is thin in academia, government, and industry because of the low levels of funding available to support research. Undersea weapons S&T is not viewed as an attractive career path, and the current knowledge-base pipeline, while able to support current program activities, would be hard pressed to support the level of activity required for development of next-generation weapon systems, which will reflect increased levels of sophistication in virtually all the critical technology areas.

Facilities and Equipment

There is no shortage of test facilities such as water tunnels and towing tanks in the United States. In fact, most facilities are used only some 40 to 60 percent of the time. At least one (at Stevens Institute) may be demolished by its parent organization, which may have alternative uses for the real estate. Facilities used in the past for supercavitating experiments are still available. There is a trend toward more use of computational fluid dynamic facilities, but the results need experimental testing. So some facilities will continue to be necessary, and the associated technical capabilities, personnel, and equipment must be of high quality. Acoustic test facilities appear to be adequate for undersea weapons S&T. There appear to be sufficient numbers of research vessels and commercial craft that can be leased, as necessary, to support an S&T program. Distributed, secure simulation facilities are needed in greater numbers and capability, and the current plans for these should be encouraged to achieve greater efficiency and economy. Otherwise, however, facilities and equipment are not in short supply.

Scope of Naval Responsibility for Undersea Weapons S&T

The Navy is the only organization in the United States that has a continuing responsibility for undersea weapons S&T. As such, its main responsibilities are to fund research and development and facilities and to set goals. Research on warheads, propulsion, sensors, control, navigation, communication, and stealth must be funded adequately to maintain the expertise needed for improving existing torpedoes and introducing innovative devices. Adequate laboratories and other facilities are necessary to support the R&D efforts. Because all completely new torpedo manufacture has been suspended for an indefinite period, the ONR undersea weapons S&T program becomes even more important.

Scope of Non-Navy Entities' Responsibility for Undersea Weapons S&T

DARPA is the only non-Navy entity in the United States that has some responsibility for supporting S&T efforts to improve the performance of undersea weapons. The scope of DARPA's efforts in this area is limited but important: it is investigating nonevolutionary solutions (e.g., networking an air-delivered torpedo with distributed sensors and communications) to problems in the field of undersea weapons. However, DARPA activities in areas such as this one generally have a time limit. Successful DARPA projects continue until such time that they are ready for transitioning to a military service.

Several government agencies also fund S&T that is relevant to undersea weapons. The National Science Foundation and the National Oceanic and Atmospheric Administration fund oceanographic research, which may indirectly support the needs of undersea weapons by providing environmental information and advancing research vehicle (including UUV) technology. There is also commercial and foreign UUV activity. While UUVs and torpedoes have much in common, generally UUVs move more slowly and are not designed to carry explosive payloads or to deal with countermeasures to their missions.

Many commercial and military technologies have been leveraged and integrated into the design, development, and construction of the Navy's undersea weapons. These technologies include computing technology, materials, robotics, fiber optics, low-drift inertial measurement units for navigation, electrochemistry (batteries and fuel cells), environmental acoustics, and computational fluid dynamics. As further technology development occurs, it must be examined, understood, and, if appropriate, adapted to the undersea environment and incorporated into the Navy's future undersea weapons. An example is the use of COTS technologies in the MK-48 and MK-54 processors.

Research in energetic materials and processes is supported both by other military services and by the Department of Energy. Although the ONR undersea weapons S&T program draws on the results of those efforts, some of the energetic materials and processes brought to undersea weapons are unique enough to require specific ONR support. For example, unlike warheads designed for attacks on land targets, some underwater warheads are designed to produce large gas bubbles that will interact dynamically with the victim hull. In this area, non-Navy programs do not have sufficient scope to support Navy needs.

The once-robust industrial independent research and development (IRAD) programs for undersea weapons have been largely eliminated. Because industry does not foresee any new torpedo programs, it has been reluctant to invest its own IRAD funds and provides no direct support for undersea weapons S&T.

In summary, undersea weapons involve special technologies, adaptation of other technologies, and unique integration of all these technologies in a challenging undersea environment. There is little or no continuing non-Navy support in the United States for this type of effort.

PROGRAM FUNDING AND TRANSITION ISSUES

Program Funding and Funding Trends

Because the committee was unable to get a clear picture from the funding data provided by ONR, it could not accurately judge how well these undersea weapons S&T programs have been funded in relation to the 1996 Department of the Navy master plan.[8] Currently planned 6.2 and 6.3 (BA2 and BA3) funding for the ONR programs is shown in Figure 2.1. Figure 2.2 shows the funding history of the ONR program including, since FY94, 6.1 and "other" (e.g., direct congressional) funding.

Based on the information provided to the committee on current actual and/or projected funding for undersea weapons S&T activities over the 5 years since the Navy master plan was issued, it would appear that the gap between the funding requested by the 1996 master plan and the funding actually received has so far meant the loss of approximately 1 year's effort over the last 3 years and that the undersea warfare advanced technology demonstration (ATD) funding is coming several years later than planned. It would also appear that program element 0602747, while funded less in the early years, has been increased in FY00 and FY01 to compensate.

The slippage raises a concern about whether adequate funding is being applied to nonevolutionary

[8]Department of the Navy. 1996. *Undersea Weapons, Vehicles, and Countermeasures: Master Plan,* The Pentagon, Washington, D.C.

S&T in the ONR budget. Based on the presentations it heard, the committee's impression was that in the area of undersea weapons, most future S&T funds will be used for upgrades to lightweight and heavyweight torpedoes in lieu of programs normally contained in the Systems Command Program Executive Office 6.3/6.4 program elements. Such funding decisions would give the Navy a short-term, evolutionary focus and allow for only minimal analysis or preparation for the future. In short, program funding is not enough to meet current requirements, nor, judging by the number of torpedoes being made by foreign countries, is it enough to offset the evolving S&T available to potential enemies. Moreover, the committee's view is that basic research funding (6.1) is much too small.

Integration with and/or Transition to Higher-Budget-Category Programs

Lack of integration of the ONR undersea weapons S&T program with (and/or transition to) higher-budget-category programs is not a concern. The ONR program is so well integrated—indeed, too well integrated—into the programs administered by PMS 404 that it has become largely evolutionary in nature. In summary, ONR undersea weapons S&T programs are too tightly integrated with programs in higher-order-budget categories, and basic and applied research that could lead to revolutionary weapons is being neglected.

SUMMARY OF ASSESSMENT

Based on the preceding discussion, the committee offers the following summary assessment of the ONR undersea weapons S&T program:

- Torpedo upgrades are mature.
- Interactions with related technology areas need strengthening.
- Program funding is not sufficient to offset the evolving S&T available to potential enemies. In particular, basic research funding (6.1) is much too small.
- Because they must function in a challenging undersea environment, undersea weapons involve special technologies, adaptations of other technologies, and unique integration of all these technologies. There is no sustained non-Navy support in the United States for this type of effort.
- The U.S. Navy must make a greater effort to provide leadership in undersea weapons research and development if it wishes to match the activity and capability of other nations.
- The knowledge-base pipeline is adequate to support the current program, although undersea weapons research is not viewed as a particularly attractive career path. However, this pipeline would be hard pressed to support the level of activity required for the development of next-generation weapon systems, which will be increasingly sophisticated in virtually all the critical technology areas.
- Facilities and equipment are not in short supply, although distributed simulation facilities in greater numbers and capability will be needed.
- The integration of the ONR undersea weapons S&T program with torpedo programs in higher-order budget categories is too tight. Basic and applied research that could lead to revolutionary weapons is being neglected.

In answer to the first key question in the terms of reference concerning what technologies are needed but are not being developed by the ONR undersea weapons S&T program, the committee offers the following judgments:

- The current approach to effectively confronting submarines in the littoral environment is not founded on a complete analysis and a good understanding of the physics of the problem, and it needs attention at the most basic level. Within the ONR undersea weapons S&T program, support for the underlying S&T is minimal.
- Deployable, distributed sensor arrays are a promising technology that needs to be built upon, as does related work in data fusion and undersea communications.
- Undersea weapons applications of fiber-optic bandwidth need to be exploited.
- UUVs and small manned underwater vehicles could be employed by naval forces as semiautonomous, long-endurance hunter/killers and reconnaissance vehicles.
- Alternative prime power concepts (e.g., hybrid advanced electric and internal combustion systems) that might be applicable to weapon-carrying and reconnaissance undersea vehicles need to be part of the exploratory program.
- Within the ONR undersea weapons S&T program, the committee received no indication of program activity in short-action-time, rocket-propelled, air- or surface-delivered undersea weapons.
- The committee did not note any programs based on other than traditional torpedo concepts.
- There is a need for the disciplined use of operations and systems analysis as a means to evaluate, quantify, and guide program decisions.

In answer to the second key question in the terms of reference, namely, the extent to which undersea weapons S&T depends on Navy-sponsored R&D, the committee believes that undersea weapons involve the development of special technologies, adaptation of these and other technologies to the undersea environment, and unique integrations of all these technologies into a weapon. In the United States the only sustained support for these kinds of efforts to develop and produce undersea weapons comes from the U.S. Navy.

3

The Future of Navy Undersea Weapons: Important Issues

The consensus of the committee is that more effort will have to be devoted to future undersea warfare operational concepts and to operations and systems analysis of the enabling technologies to validate their utility and fully understand their implications. This effort should involve a broad spectrum of Navy, academic, and industrial expertise and should lead to the criteria for and the architecture of the future major weapons development programs. In this chapter, the committee lays out what it believes are the most important issues surrounding the future of Navy undersea weapons.

UNDERSEA WARFARE AS A CONTEXT FOR ASSESSING THE UNDERSEA WEAPONS SCIENCE AND TECHNOLOGY PROGRAM

An assessment of the Navy's science and technology (S&T) program for the development of undersea weapons, much like any other assessment, cannot be conducted meaningfully without a proper context within which to frame judgments. Weapons are ultimately developed to fit within military operations designed to control enemy behavior in time of war, and the Navy chose this larger context, undersea warfare operation, as the basis for developing its current undersea weapons capability.

The collapse of the Soviet Union fundamentally changed the nature of undersea warfare. During the Cold War the challenge was posed exclusively by torpedoes fired from relatively large nuclear submarines operating in the vast expanses of the worlds' oceans; in the future the primary undersea threat is more likely to come from smaller diesel boats operating in the constrained waters of the world's littorals and will include weapons fired from a variety of surface, air, and land platforms. In addition to this undersea weapons threat, the Navy will have to contend, on the one hand, with the minefields most enemy nations could easily deploy in defense of their territorial waters and, on the other, with the diminished tolerance Americans seem to have developed for nearly any loss of human life in defense of security interests other than national survival.

Secure in the belief that, at least for the immediate future, no numerically sizable naval threat is likely to materialize and guided in part by the momentum of programs conceived during the Cold War, the Navy has not yet articulated a concept of undersea naval operations that would recognize this new paradigm. By default, the Navy expects to fight the new enemy in much the same way it intended to fight the old one and is therefore striving to adapt its old systems to the new circumstances. Thus, submarines, surface ships, and air systems are being modified to detect and localize the enemy diesel submarine in littoral environments and to destroy it with the help of conventional torpedoes. Air superiority is expected to eliminate the torpedo threat posed by enemy surface, air, and land platforms, and mine countermeasure systems currently under development are expected to forge a clear passage ahead of U.S. forces.

Absent effective guidance from the Navy leadership on likely changes in the nature of undersea warfare, the S&T community will continue to move in the direction dictated by institutional preservation and inertia and will therefore probably try to improve existing weapon systems beyond the point of diminishing returns.

Analyses conducted over the last decade have repeatedly shown that incremental improvements in endgame antisubmarine warfare (ASW) systems will not yield adequate results unless current overall capabilities are significantly improved. To make a difference for future undersea warfare needs, endgame improvements must therefore be coupled with corresponding improvements in front-end systems. In particular, torpedo improvement programs must be pursued in conjunction with programs to improve detection and localization as part of an overall systematic endeavor to improve the nation's undersea warfare capabilities.

Unfortunately, detecting and localizing small, quiet submarines operating near the ocean boundaries at near-zero speeds with conventional sonars developed to operate in the open ocean against large nuclear submarines is technically enormously difficult. Current efforts to significantly improve front-end capabilities by incremental changes to extant systems show little promise, and alternative methods better adapted to the specific circumstances of the new threat mode are not being seriously pursued. Under the circumstances, spending money and time to improve the current generation of weapons systems is of questionable benefit.

THE IMPERATIVE FOR NEW WEAPON CONCEPTS

The committee believes that there is a real penalty in pursuing the current evolutionary course of weapons upgrades instead of pursuing new concepts. The first line of argument in favor of new weapon concepts starts from the recorded performance of the current torpedoes. As indicated in recent war plan analyses, a large number of torpedoes would have to be fired to destroy an enemy submarine. The committee has not been provided with convincing evidence that the improvements being pursued under the program of record would significantly change this picture, so the required number of torpedoes is likely to remain high. Such a number could not be long sustained by the current U.S. inventory, however, if the Navy does not contemplate buying any more torpedoes. New weapon concepts must therefore be developed to get the Navy out of this dangerous predicament.

The second line of argument in favor of new weapon concepts starts with programmatics. The Navy's curtailing of torpedo acquisition could lead to a dangerous shortage of torpedoes for existing U.S. platforms and an equally dangerous loss of realistic training with torpedoes. In any case, the need for further acquisition will soon become apparent, at which time the Navy will have to consider buying additional torpedoes to maintain the inventory. Unless it has by that time developed alternative technologies for the current weapons, it will have to acquire a weapons system that would by then be obsolete. Furthermore, the nation's industrial expertise in torpedo manufacturing has atrophied and may be nonexistent when called upon.

The final line of argument has to do with the community of scientists and technologists working in undersea weapons. Absent a vigorous research program focused on developing new weapons, the community of experts will soon vanish; a program limited to product improvement is not likely to convince talented people (in either government or industry) that the Navy is serious about long-range funding for this kind of research.

THE PLANNING PROCESS

Operations and systems analysis disciplines have been expressly created to provide decision makers with an explicit statement of the critical factors that drive each decision and to allow an informed discussion about the effects of various assumptions on the decision. Whether the decision concerns a choice between systems or a choice of strategic direction, operations and systems analysis invariably helps the decision maker justify and defend his or her decision on the basis of clearly articulated reasoning and convincingly illustrated outcomes. Most important, however, the use of operations and systems analysis in support of decision making can effectively limit the influence of politically driven opinion.

Yet the Navy has now effectively replaced operations and systems analysis as the basis for making investment decisions with an integrated product team-based consensus-building process. The lack of analysis has obscured the consequences that flow from the narrowly evolutionary character of the Navy's program of record. Indeed, the consequences of placing very expensive submarines in the constrained waters of many littoral environments and of asking them to fight at a disadvantage against enemy submarines would quickly become apparent to the Navy as it analytically deduced the implications of this mode of operation.

Finally, the committee found that, influenced by a world where operations and systems analysis has no real currency, the S&T community is developing treatments for what appears to ail U.S. torpedoes without having first used analysis to diagnose the cause of the illness. Such an approach to product improvement will not lead to undersea weapons that are effective in likely future operational scenarios.

Given all these consequences of the Navy's having eliminated operations and systems analysis as part of the basis for decision making, the committee is not persuaded that the program of record will provide adequate undersea capability in the short term or that the direction in which the Office of the Chief of Naval Operations is pushing undersea weapons S&T will provide that capability in the long term.

OPERATIONS AND SYSTEMS ANALYSIS IN SCIENCE AND TECHNOLOGY PLANNING

To better align Office of Naval Research progress with the likely future, an integrating process is needed. This process would project the probable threat, devise various scenarios, and develop corresponding concepts of operations (CONOPS). For a realistic evaluation of the alternatives, it would require operations and systems analysis of options and technological possibilities. It would also require people skilled in analysis methodology and knowledgeable about system operations, enabling technology, and the use of key modeling tools.

The Navy has a long history of evaluating its undersea warfare systems using mathematical tools. During World War II, the Navy Operations Evaluation Group (OEG) and the ASW Operations Research Group were formed to analyze ASW operations.[1] OEG assessments were turned into search-and-

[1]Meigs, Montgomery. 1992. *Slide Rules and Submarines*, National Defense University Press, Washington, D.C.

engagement procedures and tactics ranging from aircraft search to screening and attack. During the early stages of the Cold War, this effort continued at a slower pace until it was reinvigorated in the mid-1960s with the establishment of the Manager ASW System Project Office, PM-4. The mission of this office was, among other things, to assess the existing and projected threat and ASW systems and to use that information to improve current fleet performance and to guide acquisition decisions. The effort continued under the sponsorship of the Chief of Naval Operations through the 1980s and culminated in the promulgation of ASW Top Level Warfare Requirements (ASW TLWR). ASW TLWR rooted systems requirements in real-world fleet performance and doctrine and based them on analyses of CONOPS that ran the gamut from short-term, limited war through full-scale, escalating war, including a strategic/theater ASW campaign, sea lines of communication protection, and combined carrier/battle group operations. These tools permitted evaluating the marginal impact of systems, system upgrades, tactics, strategy, and changes in threat.

The Navy's ability to conduct such analyses has been largely lost in the post-Cold War era, but the need for it has not diminished. Studies by the National Research Council and other review groups have repeatedly pointed out the need for these tools and their associated CONOPS.[2] Systems analyses and CONOPS should guide decisions on the allocation of Navy S&T resources. Analysts should support the integrated product team oversight groups so that investment in S&T serves the most important needs. The emerging integrated warfare architecture process should also be supported by good (and agreed-to) models.

However, good analysis entails much more than good models. It must involve military and technical experts in a process that ensures realistic estimates of performance in realistic tactical settings. The process needs independent validation and high-level (indeed all-level) support and must allow advocates and sponsors, assessors, modelers, and operators to interact so that they can extract the very best of every idea where the needs are well known and at least somewhat agreed to. There must be a mechanism for airing politically incorrect analysis.

These CONOPS and the supporting analysis should be guided by operational force plans and procedures. The analysis should permit evaluation of a range of scenarios, threat estimates, and levels of allied participation, as well as of tactics, doctrine, rules of engagement, system performance, and so on. These can easily be bounded to enhance cogent decision making. Right now, littoral warfare is a nebulous concept that leads to equally nebulous rationales for force and system planning. The committee believes that more definitive planning is essential and that such planning must be supported by high-quality operations and systems analysis.

LEAD TIME FOR NEW TORPEDOES

In 1995 the Department of Defense, in concert with the Navy, made the decision to phase out torpedo production, saying that there was no foreseen need and that the industrial base could be reconstituted quickly. In terms of programming, that decision is 10 years old. Development of a completely new torpedo is at the least a 15-year process (probably more like 20 years, given the defunct status of the

[2]Naval Studies Board, National Research Council. 1997. *Technology for the United States Navy and Marine Corps, 2000-2035, Vols. 5 and 7,* National Academy Press, Washington, D.C.; Naval Studies Board, National Research Council. 1993-1994. *Mine Countermeasures Technology, Vols. 1-4,* National Academy Press, Washington, D.C.; Naval Research Advisory Committee. To be published. *Report of the Unmanned Vehicles (UV) in Mine Countermeasures,* draft, Arlington, Va.

industrial base for making new torpedoes and the Navy acquisition bureaucracy; see Box 1.1).[3] If the Navy is to begin gearing up for new torpedo development, or more appropriately, for undersea weapons development, it should be starting a major technology assessment and development effort within the present Future Year Defense Program. There are no signs, however, that such an effort is forthcoming.

The country could lose its long-held competence in torpedoes. This state of affairs is the result of several factors:

- The absence of new torpedo development programs;
- A reduction in inventories;
- Minimal or inadequate fleet firings and in-service tests;
- Proliferation of offshore torpedo technologies, including countermeasures, that are approaching par with existing U.S. technologies;
- Export prohibitions that preclude U.S. industries from competing for international business; and
- Inadequate investment in S&T against future need.

Industry, which was such a strong player in the past, no longer has an incentive to maintain its competence or infrastructure or to invest in its own research and development (R&D). Funds for 6.4, once robust, are no longer available to stimulate and advance technology. Perishable human expertise and know-how in this unique weapons field are rapidly disappearing. Additionally, the prospect of being unable to compete in the international arms market because of draconian export constraints is a disincentive for industry. Industry is further removed as a participant since most Navy S&T funds are spent in-house.

Inventory reductions are leading to a "platinum bullet" syndrome: the belief that torpedoes are too rare and precious to waste on low-confidence targets, which will as a result linger as clutter in the battle space unless classified and eliminated. Torpedo firings during fleet exercises are dangerously low (<2 percent of inventory), and the inventory is not being adequately tested or cycled. The need to thoroughly test torpedoes is well established: inadequate testing of torpedoes before World War II resulted in weapons that failed in combat, with catastrophic consequences. The committee is concerned that history may repeat itself.

The end of the Cold War has unfettered European Community and former Soviet bloc arms manufacturers. They are now offering torpedoes, countermeasures, and undersea warfare systems to any buyer. These technologies (at least on paper) appear comparable to or better than anything the United States could now offer. At the present rate of S&T investment (the only source of undersea weapons upgrades), U.S. systems will fall behind. The prospect of this situation makes S&T investment more urgent than ever. Funding for Navy S&T is all that maintains U.S. competence in undersea weapons, and it is inadequate.

REVITALIZING UNDERSEA WEAPONS DEVELOPMENT

The Navy should reevaluate the general direction of its current undersea weapons program and provide adequate funding for the resulting new program. As a starting point for this reevaluation, the

[3]The MK-54 hybrid torpedo is now in low rate initial production and is scheduled to start production in FY02, about 7 years from its program start in FY95. It is not, however, a completely new torpedo. Most of its components come from the MK-50, MK-46, and MK-68 systems.

Navy should resurrect the operations and systems analysis capability it once used to guide decision making and use it again to determine which concepts of operation would be most cost-effective within the context of the likely future warfighting scenarios. The concepts of operation that emerge from such analysis should then serve as a foundation for additional analytic work aimed at deciding which technology would best serve the Navy's undersea weapons development.

The Navy should reconsider its position and reassess its risks against the likely level of undersea threat to its surface forces. As is done in other areas, the risk should be taken, roughly, as the product of the probability of failure and the consequence of a loss. This reassessment will not be easy and will require the intervention of highly placed military leaders; the "hammer" such leaders could wield could do much to change the current laissez-faire attitude about the undersea threat. The recent impetus the submarine sonar program received from ADM Bruce DeMars, USN (now retired), and the highly successful ARCI program that resulted from that intervention (discussed in Chapter 2 under "Weapons Design Optimization" and in Appendix B), is an excellent case in point.

Next, the Navy should take advantage of the heightened concern about the undersea threat that would come from such a push and try to provide adequate funding. The funds should not only support the enlarged research program that would emerge from the concept studies mentioned above but would also have to underwrite the beginning of a future acquisition program. As has already been explained, an acquisition program is long overdue. First, since it has taken 15 to 20 years to develop a completely new generation of torpedoes, there may not be much time to update U.S. undersea warfare capability. Second, the Navy should quickly revive the incentives for industrial investment in R&D, which can be brought about only by an acquisition program. And finally, the Navy should provide the exciting research activity that can attract young scientific and technological talent, without which the nation will not be able to maintain a credible undersea warfare capability into the next century.

No other organization will do these things for the Navy.

4

Findings, Conclusions, and Recommendations

Based on the discussion in the preceding chapters, the committee offers the following four sets of findings, conclusions, and recommendations:

Finding 1. Undersea weapons involve special technologies, adaptations of other technologies, and unique integration of all these technologies for which there is in the United States little non-Navy-sponsored research and development (R&D). The committee found some good examples of ongoing healthy and productive S&T, including the following:

- The program on propulsion at the Applied Research Laboratory, Pennsylvania State University (ARL/PSU) is exemplary and offers technologies for both weapons and vehicles that could be used in future systems. Closed-cycle engines are among the increasingly attractive options as the importance of stealth and endurance increases.
- The programs on warheads at the Naval Surface Warfare Center, Indian Head are good examples of R&D in a mature area that has consistently delivered fresh results in S&T and new generations of explosive compounds tailored to the Navy's needs. Current research on the penetration of hardened hulls is important. Research on the problems of sensitivity of high-energy materials should be supported.
- The program at ARL/PSU and the Naval Undersea Warfare Center to develop a high-speed, supercavitating vehicle is challenging and sufficiently promising to warrant research in (1) the physics of supercavitating flow, (2) vehicle control and guidance methodology, and (3) the design and building of a testbed. There should be careful analysis of the operational utility of the concepts this technology could include.

On the other hand, the committee believes that a truly healthy undersea weapons S&T program should include industry participation, but industry is not now a significant participant or investor in undersea weapons S&T.

Conclusion 1. The Navy has a unique responsibility for the support and health of S&T related to undersea weapons construed broadly. Although the information presented to the committee pertained only to torpedo-related matters, the committee believes that its recommendation on this responsibility should apply to all weapons spending some time undersea. The undersea weapons S&T effort should include industrial participation, at least in the relevant concept definition studies, and related operations and systems analysis.

Recommendation 1. The Navy should designate S&T for undersea weapons—construed broadly—as a National Naval Need. Because of the key enabling characteristics of undersea weapons for the fleet and the need for industry involvement, the Navy should also consider designating undersea weapons as one of the Future Naval Capabilities, a step that would allow it to begin preparation for a new weapons acquisition program.

Finding 2. There is no broadly based, future-oriented program of operations and systems analysis in place to support Office of Naval Research (ONR) S&T planning in undersea weapons. With regard to the individual ONR undersea weapons program areas, the committee found the following:

- It is not obvious that the programs on guidance and control at ARL/PSU and NUWC are succeeding at coping with progressively quieter targets and evolving countermeasures. The careful operations and systems analysis needed to critically assess operational performance in matters of target detection, identification, and homing seems to be missing.
- Upgrades intended to quiet the MK-48 and MK-54 torpedoes (mainly by NUWC) were not persuasively presented to the committee. The open-cycle engine, buoyancy disadvantages, hydroacoustic noise, and other characteristics make the upgrades questionable in light of the evolving stealth and countermeasure capabilities of potential enemy targets. No systems analyses of predicted program success or time scales for acquisition were presented to the committee.
- A number of plausible approaches to defending against torpedoes were broadly outlined to the committee, including noisemakers, decoys, supercavitating pellets, and antitorpedo torpedoes. Individually these might be of value, but maximum benefit will be achieved only if they are integrated properly into a plausible, coherent defense architecture system.
- Weapons design optimization, which appears to be a relatively recently identified effort, while useful still does not satisfy the need for operations and systems analysis called for at several points in this report.

Conclusion 2. Concept definitions, and systems and operational analysis, are needed in a number of program areas and as a part of a healthy and productive S&T process generally.

Recommendation 2. ONR should rigorously implement a process of operations and systems analysis of undersea weapons systems. Operational performance in both littoral and blue water environments should be covered. Emphasis should be placed on enabling science and technology and weapons systems of advanced mission and design.

Finding 3. The health of the existing Navy program on undersea weapons S&T is strongly affected by the present emphasis on upgrades of existing torpedo systems. Less than 10 percent of the ONR undersea weapons S&T budget was for basic research (6.1) in 1999. The health of the program could be improved by much greater attention to S&T issues that will affect future weapons systems.

Conclusion 3. To be more forward-looking, greater S&T emphasis at a fundamental (6.1/6.2) level is needed within the ONR undersea weapons effort.

Recommendation 3. ONR should increase undersea weapons S&T funding sufficiently to satisfy Future Naval Capability goals while ensuring that longer-range, higher-risk/higher-payoff alternative S&T is also enabled. This investment in future systems must be protected against raids to bail out near-term projects.

Finding 4. While some of the items covered in the present program, which is focused on upgrades of existing torpedoes, may also be useful in future weapons systems (e.g., propulsion units and warheads), long-range exploration for fundamentally new undersea weapons concepts and missions is needed.

Conclusion 4. Innovation beyond current undersea weapons concepts and missions is needed for a healthy S&T program. S&T should be pursued toward torpedoes operating with sensor arrays, unmanned and manned undersea vehicles (attack, reconnaissance, and so forth), and sophisticated mines, and toward achievement of short-action-time air-delivered undersea weapons systems. In the future, undersea weapons systems will be driven increasingly by overall architecture that will demand much more interdisciplinary coordination than was seen by the committee in the course of its study.

Recommendation 4. ONR should take a broader and longer-range view of undersea weapons systems, specifically those not limited to torpedoes and beyond the 5-year horizon. Some such concepts are being explored in other ONR undersea technology activities and in the Defense Advanced Research Projects Agency and should be considered.

Appendixes

A

Technology Insertion Road Map

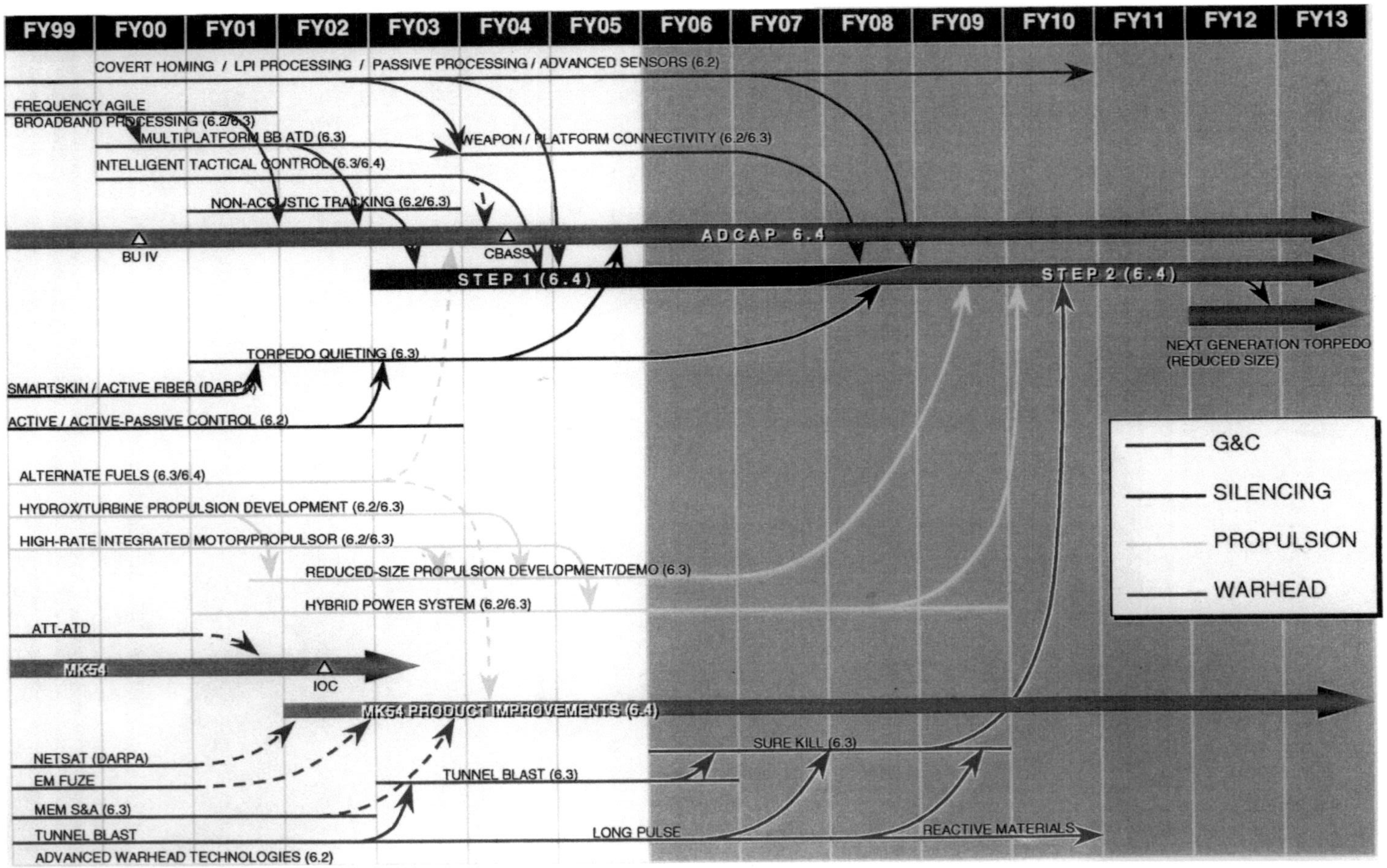

FIGURE A.1 Technology insertion road map showing the expected schedule, according to the Undersea Weapons, Vehicles, and Countermeasures Master Plan, for technology improvements coming from the four R&D programs (guidance and control, silencing, propulsion, and warhead) and also from DARPA, to be inserted in the planned evolution of the U.S. Navy's heavyweight torpedoes (MK 48, top section of the road map) and lightweight hybrid torpedoes (MK 54, lower section). The map also indicates the schedule and type of funding for the R&D programs (6.2 and 6.3) and engineering development (6.4), as well as the schedule for the advanced technology demonstration (ATD) of the MK 54 as an antitorpedo torpedo (ATT) countermeasure for surface ships. Acronyms are defined in Appendix D.

SOURCE: Timothy Douglass, Program Executive Officer for Undersea Warfare, "Torpedo S&T: An Emerging Crisis," briefing to the committee, August 31, 1999.

B

Lessons of the Advanced Rapid COTS Insertion Process

The committee was impressed by the significant change in process achieved by the submarine sonar community that allowed advanced rapid commercial-off-the-shelf insertion (ARCI) in all submarines. It believes that three important lessons were learned from that experience.

- *Lesson 1: Operational testing must be adequate and must be carried out under realistic conditions.* The change was motivated by real-world operational experience suggesting that U.S. submarine sonars were not performing up to expectations. A similar situation is believed to exist with regard to undersea weapons performance based on exercises and simulations. One of the basic tenets introduced into the ARCI process was the extensive use of at-sea data for several purposes: (1) laboratory evaluation of specific algorithms; (2) end-to-end laboratory testing of system upgrades or builds; (3) at-sea testing prior to introduction into the fleet; and (4) feedback from at-sea operations of certified systems.
- *Lesson 2: Degraded performance must be understood at a fundamental level.* The increased attention paid to a first-principles understanding of sonar and validations through testing with real data was one of the key factors in the ARCI process. This special process of build-test-analyze-understand-build-test allows new builds to be introduced quickly into the fleet with very short development cycles for most software algorithms. While testing for weapons system science and technology may not need to be so extensive, the Office of Naval Research could incorporate the lessons of the ARCI process as appropriate.
- *Lesson 3: When senior Navy management became involved, bureaucratic barriers fell.* In the ARCI experience, the concerns expressed by Admiral DeMars galvanized a multifunctional, expedited approach. A broad-based data-gathering, analysis, and implementation program was successfully and rapidly implemented and appears to have yielded excellent results. In other words, pressure from senior Navy leadership was applied and overcame bureaucratic obstacles to progress.

C

Biographies of Committee Members

Albert J. Baciocco, Jr., Chair, retired from the U.S. Navy as Vice Admiral in 1987 after 34 years of distinguished service, principally in the nuclear submarine force and directing the Department of the Navy research and technology development enterprise. He graduated from the U.S. Naval Academy in 1953, receiving a B.S. in engineering, and subsequently pursued graduate level studies in nuclear engineering as part of his training for the U.S. Navy nuclear propulsion program. In 1998, Admiral Baciocco received an honorary doctorate in engineering from Florida Atlantic University. Upon retirement from the Navy, Admiral Baciocco formed The Baciocco Group, Inc., a technical and management consulting practice, and has since been engaged in a variety of business and pro bono activities with industry, government, and academe, most of them related to technology planning, investment, management, and implementation. Admiral Baciocco is a director of the American Superconductor Corporation. He serves on several boards and committees of government, industry, and academe. He is a member of the Naval Studies Board and of the Board of Trustees of the South Carolina Research Authority, a state-chartered, not-for-profit advanced technology and management enterprise. In addition, he is a director of its affiliated Advanced Technology Institute. Admiral Baciocco serves on the boards of directors of the University of South Carolina Research Institute, the Foundation for Research Development at the Medical University of South Carolina, and the Institute for Ocean and Systems Engineering at Florida Atlantic University. He currently serves as chair of the Southeastern Universities Research Association's Maritime Technical Advisory Committee to the Thomas Jefferson National Accelerator Laboratory in Newport News, Virginia. In addition, he is a member emeritus and past chairman of the Advisory Board to the Applied Physics Laboratory, University of Washington, and he is currently a member of the Naval Studies Board.

Arthur B. Baggeroer is Ford Professor of Engineering and Secretary of the Navy/Chief of Naval Operations Chair for Ocean Science in the Departments of Ocean and Electrical Engineering at the Massachusetts Institute of Technology. He is a member of the National Academy of Engineering and a fellow of the Institute of Electrical and Electronic Engineering and the Acoustical Society of America. He has been the director of the Massachusetts Institute of Technology (MIT)-Woods Hole Joint Program in Oceanography and Oceanographic Engineering. During sabbatical leaves he was a consultant to the Chief of Naval Research at the Supreme Allied Commander, Atlantic (SACLANT) Center in La Spezia, Italy, and a Green Scholar at the Scripps Institution of Oceanography. Dr. Baggeroer's research is primarily related to advanced signal processing methods applied to sonar, ocean acoustics, and geophysics.

Alan Berman, an independent consultant, currently consults for the Center for Naval Analyses, where he assists with analyses of Navy research and development (R&D) investment programs, space operation capabilities, and information operations. He also consults for the Applied Research Laboratory of Pennsylvania State University, where he provides general management support and program appraisal. Dr. Berman's background is in defense research and technology, particularly in regard to advanced weapon and combat systems. He is regarded as a leading expert on combat systems. He was at one time dean of the Rosenstiel School of Marine and Atmospheric Sciences at the University of Miami and at another, director of research at the Naval Research Laboratory. Dr. Berman has served on numerous government advisory and scientific boards. He is currently a member of the Naval Studies Board. He is also a member of the Free Electron Laser (FEL) oversight board that advises the Thomas Jefferson National Accelerator Laboratory of the Department of Energy on its FEL program.

Gerald A. Cann, an independent consultant, currently consults for the Raytheon Company. He recently retired as senior advisor to the Raytheon Executive Office. Mr. Cann's background is in system development from the viewpoints of both industry and government. Mr. Cann is a former vice president of the General Dynamics Company, where he created a new business unit that formed alliances with both internal divisions of the company and major outside participants in undersea warfare. He has held several government positions, including Assistant Secretary of the Navy for Research, Development, and Acquisition (responsible for acquisition policy, procedure, and execution of all research, development, production, shipbuilding, and logistics programs); Principal Deputy Assistant Secretary of the Navy for Research, Engineering, and Systems; and Director of Naval Warfare in the Department of Defense for Research and Engineering. He completed his term as a member of the Naval Studies Board in 1999.

A. Douglas Carmichael is professor (emeritus) of power engineering at the Massachusetts Institute of Technology. His background is in naval propulsion systems. He joined the Department of Ocean Engineering at MIT in 1970, where submarine propulsion was his primary research interest. Prior to joining MIT, Dr. Carmichael was research fellow at the Imperial College of Science and Technology, London, England. He is the author of a very large number of publications on naval propulsion, including the design impacts of alternative technologies. He is a fellow of the Society of Naval Architects and Marine Engineers.

Thomas A. Clare is a senior consultant to Techmatics, Inc., an Anteon Company. He has more than 30 years of diversified management and senior executive experience in the research and development of large, complex systems, software development, science and technology management, human resources management, and R&D laboratory management. As the former executive director of the Navy's largest R&D laboratory, Dr. Clare made major contributions to the research and development of undersea weapons systems, surface combatants, aircraft carriers, Navy and Marine Corps systems, including Aegis, the submarine-launched ballistic missile Tomahawk, and theater ballistic missile defense weapon and combat systems. He has also been a key advisor and leader in the Navy and the Department of Defense scientific and technology community, particularly in systems engineering and weapon system concepts. He has served on numerous government advisory and scientific boards, addressing the need for systems engineering and management policies and practices across the services in an era of reduced budgets. He is a member of the American Society of Naval Engineers, the Surface Navy Association, and Mary Washington College President's Roundtable.

Dennis F. Colin is vice president and general manager of Navigation and Gravity Systems at Lockheed Martin Systems. His expertise is in the integration of complex electronics systems on submarines, integrated undersea surveillance systems, and surface, and air platforms. At Lockheed Martin, Mr. Colin has responsibility for all fleet support, training systems, equipment design and procurement, and program management for Trident Navigation and Gravity Systems. Prior to this assignment, he had overall responsibility for Lockheed Martin's Navy submarine programs, as well as for interfacing with Lockheed Martin facilities dealing with the coordination of U.S. Navy nonpropulsion requirements on submarines. Mr. Colin is a member of many professional organizations, including the Navy League.

Alec D. Gallimore is associate professor of aerospace engineering and director of the Plasmadynamics and Electric Propulsion Laboratory at the University of Michigan. His background is in electric propulsion. An experimental plasma physicist, his current research interests include electric propulsion engine diagnostic tools for plume characterization. Dr. Gallimore is an associate fellow of the American Institute of Astronautics and Aeronautics and a member of the American Society for Engineering Education. He recently served on the Defense Science Board's Task Force on Force Modernization.

Ernest L. Holmboe currently works for the Applied Physics Laboratory/Johns Hopkins University (APL/JHU) as a part-time special assistant. His background is in systems engineering, particularly in connection with submarine technology. Until June 1998, Dr. Holmboe served as the head of the Submarine Technology Department (STD). The STD concentrates on submarine security and undersea surveillance and is also applying its technical knowledge to such projects as special operations and technologies to reduce manpower on ships and submarines. Before joining APL/JHU, Dr. Holmboe was head of AT&T Bell Laboratories' Navy Systems Engineering Department. He is a member of the U.S. Naval Institute, the Naval Submarine League, and the National Defense Industrial Association.

Alfred I. Kaufman is a research staff member for the System Evaluation Division at the Institute for Defense Analyses (IDA). His background is in undersea warfare, particularly acoustic detection. A physicist by training, Dr. Kaufman's research interests include the analyses of sonar and acoustic processing systems, as well as wave propagation in turbulent mediums. He was also a leading expert on national security issues relating to the Soviet threat to NATO sea lines of communication. Before joining IDA, Dr. Kaufman was an analyst at the Center for Naval Analyses and head of the Physics Department at Manhattanville College. He is a member of the American Physical Society.

David W. McCall is a retired director of the Chemical Research Laboratory at AT&T Bell Laboratories (now Bell Laboratories, Lucent Technologies). A member of the National Academy of Engineering, he has a background in materials science and chemistry. As director of the Chemical Research Laboratory at AT&T Bell Laboratories, he was responsible for research, development, and engineering of materials and processes employed in the production of communications systems. The materials included a variety of insulating materials and advanced structural materials that were essential to the long-term integrity of the equipment produced. During his career at AT&T Bell Laboratories, Dr. McCall served in a number of capacities, including head of the Physical Chemistry Department. There he led programs expanding the bandwidth of coaxial submarine cable systems and programs to advance the understanding of atmospheric chemistry as it relates to detection and tracking of intercontinental ballistic missiles. Dr. McCall has served on numerous government advisory committees and was appointed by the President as chair of the National Commission on Superconductivity. In 1999 he completed three terms as a member of the Naval Studies Board.

L. David Montague, an independent consultant, is retired president of the Missile Systems Division at Lockheed Martin Missiles and Space. A member of the National Academy of Engineering, he has a background in military weapon systems, particularly the guidance and control of submarine-launched weapons. In addition to his expertise in both tactical and strategic strike systems, his experience has focused on the requirements, development, and policy issues of defense systems to protect against weapons of mass destruction. Mr. Montague is a fellow of the American Institute of Aeronautics and Astronautics and has served on numerous government and scientific advisory boards, including task forces for both the U.S. Army Science Board and the Defense Science Board. He is a former member of the Steering Task Group of the Navy's Directorate for Strategic Systems Programs.

Douglas R. Mook is director of Advanced Systems at Sanders. His background is in acoustic processing and sensor fusion. Trained in both electrical and ocean engineering, Dr. Mook is responsible for several key Department of Defense programs, including the U.S. Navy's Advanced Acoustics Communications Advanced Technology Demonstration, the U.S. Army's Federated Laboratories digital battlefield programs for communications and sensors, and the Defense Advanced Research Projects Agency's unattended ground sensors programs. He is a member of the Navy's fleet ballistic missile submarine security review committee and has been a member of the Army Research Laboratory Restructuring Committee and the Army Digital Battlefield Definition Committee.

Richard F. Pittenger, retired from the Navy as a Rear Admiral and is currently associate director for marine operations at Woods Hole Oceanographic Institution (WHOI). Admiral Pittenger's background is in antisubmarine warfare and oceanography. During his naval career, he was oceanographer of the Navy and also director of the Antisubmarine Warfare Program. At WHOI, Admiral Pittenger oversees the operations of three large research ships, the WHOI research submarine, and remotely operated vehicles.

Daniel Savitsky is professor emeritus of ocean engineering and retired director of the Davidson Laboratory at the Stevens Institute of Technology. His background is in naval architecture, particularly in hydromechanics. Throughout his career, Dr. Savitsky has been a leader in research focused on hydrodynamics, as well as other ocean engineering concerns associated with naval architecture design. He is a fellow and honorary member of the Society of Naval Architects and Marine Engineers and a member of the American Society of Naval Engineers.

D

Acronyms and Abbreviations

ADCAP	advanced capability (MK 48) torpedo
ADM	advanced development model
ARCI	advanced rapid COTS insertion (process; also program)
ARL/PSU	Applied Research Laboratory, Pennsylvania State University
ASROC	antisubmarine rocket
ASW	antisubmarine warfare
ATD	advanced technology demonstration
ATT	antitorpedo torpedo
AUR	all up round
BB	broadband
CBASS	Common Broadband Advanced Sonar System
CIWS	close-in weapon system
CM	countermeasures
CNO	Chief of Naval Operations
CONOPS	concept of operations
COTS	commercial off the shelf
DARPA	Defense Advanced Research Projects Agency
DOD	Department of Defense
EM	electromagnetic
FNC	Future Naval Capability
G&C	guidance and control

HWT	heavyweight torpedo
IMU	inertial measurement unit
INS	inertial navigation system
IOC	initial operating capability
IRAD	independent research and development
LELFAS	low-endurance low-frequency active surveillance
LHT	lightweight torpedo
LPI	low probability of intercept
LRIP	low rate initial production
LWT	lightweight hybrid torpedo
MEM	microelectromechanical
MODS	modifications
MSII	milestone two
NETSAT	networked sensors and torpedoes
NRC	National Research Council
NSWC/CD	Naval Surface Warfare Center, Carderock Division
NSWC/IH	Naval Surface Warfare Center, Indian Head
NUWC	Naval Undersea Warfare Center
OEG	Operations Evaluation Group
O&MN	operations and maintenance, Navy
ONR	Office of Naval Research
OSD	Office of the Secretary of Defense
OTIIB	part "B" of the operational evaluation (conducted prior to the Milestone III decision)
PEO-USW	Program Executive Office, Undersea Weapons
PMS	Program Manager, Naval Sea Systems Command
R&D	research and development
RDT&E	research, development, testing, and evaluation
RMS	Remote Minehunting System
S&T	science and technology
SCEPS	Stored Chemical Energy Propulsion System
SNR	signal-to-noise ratio
SSN	nuclear-powered submarine
STEP	Stealth Torpedo Enhancement Program
TLWR	Top Level Warfare Requirements
TSE	tactical support element

USV	unmanned surface vehicle
UUV	unmanned underwater vehicle
VLA	vertical launch ASROC
VTOL	vertical take-off and landing
WDO	weapons design optimization